Causal Analysis for Climate Study

This book offers the theory of causal analysis and its applications. The authors have developed this book in relation to its applications to four climatological phenomena to prove the theory of causal analysis in time sequential data analysis.

Local Causal Test and the Partial Causal Test are used to study the theory of causal analysis. These are then applied to understand the climate effects of the eruption at Mt. Pinatubo, the effect of the El Niño and Southern Oscillation (ENSO), and the North Atlantic Oscillation (NAO). The reader learns about Test(S), a statistical test used to determine if the statistical properties of the data, such as mean and variance, remain constant over time within a local window. The authors also use the Stationary Test and Causality Test in time series to explore relationships within a localized subset of data as witnessed from the climate phenomenon. The book looks at the eruption at Mt. Pinatubo, ENSO, and NAO and applies causal theory to study the resultant air temperature at the 1000 hPa pressure level, geopotential height of 1000 hPa pressure surface, and surface precipitation. The program code of the causal analysis is offered for readers to be able to reconfirm the results and apply it to other time-sequential data.

This book is useful for researchers and graduate students of applied mathematics, physics, geophysics, meteorology, oceanography, engineering technology, and economics.

Yuji Nakano is Emeritus Professor at Shiga University, with the Faculty of Economics, where he has made significant contributions to the field of economics. His most recent research was on the nonlinear causal analysis of economic time series.

Osamu Morita is currently a guest Professor at Fukuoka University, who specializes in meteorology, physics, and earth and planetary sciences. He is particularly known for his work in geophysical fluid dynamics, the processes governing the movement of fluids on Earth, such as oceans and the atmosphere.

Causal Analysis for Climate Study

Theory and Applications

Yuji Nakano and Osamu Morita

CRC Press is an imprint of the
Taylor & Francis Group, an **informa** business
A CHAPMAN & HALL BOOK

First edition published 2025
by CRC Press
2385 NW Executive Center Drive, Suite 320, Boca Raton FL 33431

and by CRC Press
4 Park Square, Milton Park, Abingdon, Oxon, OX14 4RN

CRC Press is an imprint of Taylor & Francis Group, LLC

ISBN: 978-1-032-99304-1 (hbk)
ISBN: 978-1-032-99310-2 (pbk)
ISBN: 978-1-003-60342-9 (ebk)

DOI: 10.1201/9781003603429

Typeset in CMR10
by KnowledgeWorks Global Ltd.

Contents

Preface

The purpose of this book is to introduce the theory of causal analysis together with its climate applications. The causal analysis is only known in the small community of statistical mathematicians, but it is very useful and powerful tool for time sequential data analysis. The authors intend to prove this fact in the book. The book is divided into two parts, part I (chapters 1 and 2) for the theory of causal analysis and part II (chapters 3, 4, 5 and 6) for its applications.
In chapter 1 Test(S) is proposed, which is the test whether the given data are a realization of a local and weakly stationary process or not. The theory of KM_2O-Langevin equation, which plays a crucial role in Test(S), is discussed. In chapter 2, the local causal value is defined, which is the measure of influence of stationary time series $\mathcal{Y}$ to $\mathcal{X}$. Further, the method to remove the influence of the third time series $\mathcal{Z}$ from $\mathcal{X}$ and $\mathcal{Y}$ is proposed. Chapter 3 discusses the effect of the Mt. Pinatubo eruption in 1991 on the global climate. The causal analysis and the correlation analysis are used in this chapter, and it will be shown that the cooperative effect of the large volcanic eruption and the ENSO warm event is crucial to the global climate. Chapter 4 gives the effect of the El Niño and Southern Oscillation (ENSO) on the global climate elements, the air temperature at 1000 hPa level, the geopotential height of 1000 hPa surface, and the surface precipitation, using the causal analysis. Chapter 5 offers the effect of El Niño and Southern Oscillation (ENSO) on the Japanese climate elements, the surface air temperature, the sea level pressure, and the surface precipitation, using the causal analysis. In chapter 6, we will discuss the effect of the North Atlantic Oscillation (NAO) on the global climate elements. Finally, the Appendix offers the program code of the causal analysis to readers, who are able to reconfirm the results and/or apply it to another time-sequential data analysis.
This book is suitable for undergraduate students, graduate students, and researchers of physics, geophysics, meteorology, climatology, oceanography, chemistry, and economics.

Acknowledgments

The authors heartily acknowledge Dr. Yasunori Okabe, Emeritus Professor of University of Tokyo, for his advice to construct the theory of causal analysis. They thank Dr. Paul Wessel and Dr. Walter H. F. Smith, University of Hawaii, for free use of the Generic Mapping Tools (GMT), which is used for most figures of this book. Special thanks are due to Mr. Gerald Bok, Commissioning Editor at Taylor & Francis, for Computer Science and Statistics, and Mr. Tony Moore, the senior editor of Taylor & Francis Group Ltd. for their great help to publish this book.
Thanks are due to the Physical Sciences, Laboratory (PSL), NOAA, Boulder, Colorado, USA, for providing NCEP/NCAR Reanalysis 1 data, the NASA Langley Research Center for providing SAGE II data, the Japan Meteorological Agency for free use of the precipitation data, Climatic Research Unit, University of East Anglia for providing SOI data, and Dr. Phil Arkin, Dr. Pingping Xie, and National Center for Atmospheric Research Staff for the permission of the use of CMAP data.

Yuji Nakano and Osamu Morita

Part I

Theory

Chapter 1

Stationarity Test in Time Series

by Yuji Nakano

We propose Test(S) to judge whether given data are a realization of a local and weakly stationary process or not. We overview the theory of KM_2O-Langevin equations, which plays a crucial role in constructing Test(S).

1.1 Introduction

A weakly stationary process, whose time parameter space is a finite interval of $\mathbf{T}$, is called a local and weakly stationary process. Okabe [1988] established the theory of KM_2O-Langevin equations[1] that described the framework of this theory. Test(S) is one of the results of this theory, which states a criterion that multi-dimensional data are a realization of a local and weakly stationary process [Okabe and Nakano, 1991].
We apply this Test(S) to the given data. If the data pass Test(S), viz. that they are accepted to be a realization of a local and weakly stationary time series, we proceed to further analysis. In section 1.2, we briefly overview the theory of KM_2O-Langevin equations. We summarize the deduced process of Test(S) in section 1.3.

1.2 KM_2O-Langevin Equations

Following the notation and terminology of Okabe-Nakano [1991] and Nakano [1995], we overview it in this section.

[1]Capital letters of the second names of R. Kubo (physicist), H. Mori (physicist), T. Miyoshi (mathematician), and Y. Okabe

DOI:10.1201/9781003603429-1

Let $d, N \in \mathbf{N}$. Let $\mathbf{X} = (X(n); |n| \leq N)$ be any d-dimensional stochastic process on a probability space $(\Omega, \mathcal{B}, \mathcal{P})$.

Definition 1.2.1 $\mathbf{X}$ is called a local and weakly stationary process with covariance function R if it holds that for any $n, m \in \mathbf{Z}, |n| \leq N, |m| \leq N$,

$$E(X) = E[X(n)] = \mu, \tag{1.1}$$

$$E[(X(n) - \mu)^t (X(m) - \mu)] = R(n - m). \tag{1.2}$$

Without losing generality, we assume that $\mu = \mathbf{0}$.
For any $n \in \{1, \ldots, N\}$, we define a block Toeplitz matrix $S_n \in M(nd; \mathbf{R})$ by

$$S_n = \begin{pmatrix} R(0) & R(1) & \cdots & R(n-1) \\ {}^tR(1) & R(0) & \cdots & R(n-2) \\ \vdots & \vdots & \ddots & \vdots \\ {}^tR(n-2) & {}^tR(n-3) & \cdots & R(1) \\ {}^tR(n-1) & {}^tR(n-2) & \cdots & R(0) \end{pmatrix}. \tag{1.3}$$

We assume

(A-1) $\qquad R(0) \in GL(d; \mathbf{R})$,

and

(A-2) $\qquad S_n \in GL(nd; \mathbf{R})$.

We set

$$X(n) = \begin{pmatrix} X_1(n) \\ X_2(n) \\ \vdots \\ X_d(n) \end{pmatrix} \qquad (|n| \leq N). \tag{1.4}$$

For $n_1 < n_2$, $n_1, n_2 \in \{-N, \ldots, N\}$, we define $\mathbf{M}_{n_1}^{n_2}(\mathbf{X})$, which is a closed linear subspace of $L^2(\Omega, \mathcal{B}, P)$ by

$$\begin{aligned} \mathbf{M}_{n_1}^{n_2}(\mathbf{X}) = & \text{ the closed linear hull of} \\ & \{X_j(m); 1 \leq j \leq d,\ n_1 \leq m \leq n_2\}. \end{aligned} \tag{1.5}$$

Especially, we define that

$$\mathbf{M}_0^{-1}(\mathbf{X}) = \mathbf{M}_1^0(\mathbf{X}) = \mathbf{0}. \tag{1.6}$$

For $n \in \{0, \cdots, N\}$, $P_{\mathbf{M}_0^{n-1}(\mathbf{X})}$ is a projection operator on $\mathbf{M}_0^{n-1}(\mathbf{X})$, and $P_{\mathbf{M}_{-n+1}^0(\mathbf{X})}$ projection operator on $\mathbf{M}_{-n+1}^0(\mathbf{X})$.

Now, the random forces of $\mathbf{X}$, which we call, $\nu_+ = (\nu_+(n); 0 \leq n \leq N)$, $\nu_- = (\nu_-(-n); 0 \leq n \leq N)$ are introduced:

$$\nu_+(n) = X(n) - P_{\mathbf{M}_0^{n-1}(\mathbf{X})}X(n), \tag{1.7}$$

$$\nu_-(-n) = X(-n) - P_{\mathbf{M}_{-n+1}^0(\mathbf{X})}X(-n). \tag{1.8}$$

It holds that

$$\nu_+(0) = \nu_-(0) = X(0). \tag{1.9}$$

$P_{\mathbf{M}_0^{n-1}(\mathbf{X})}X(n)$ and $P_{\mathbf{M}_{-n+1}^0(\mathbf{X})}X(-n)$ are called fluctuation parts of $\mathbf{X}$.
For any $n \in \{1, \ldots, N\}$, $k \in \{0, \ldots, n-1\}$, there exist $\gamma_+(n,k)$, $\gamma_-(n,k) \in M(d; \mathbf{R})$, such that

$$P_{\mathbf{M}_0^{n-1}(\mathbf{X})}X(n) = -\sum_{k=0}^{n-1} \gamma_+(n,k)X(k), \tag{1.10}$$

$$P_{\mathbf{M}_{-n+1}^0(\mathbf{X})}X(-n) = -\sum_{k=0}^{n-1} \gamma_-(n,k)X(-k). \tag{1.11}$$

We have

$$\mathbf{M}_0^n(\mathbf{X}) = \mathbf{M}_0^n(\circ_+), \tag{1.12}$$

$$\mathbf{M}_{-n}^0(\mathbf{X}) = \mathbf{M}_{-n}^0(\circ_-). \tag{1.13}$$

The following theorem is known as an expression formula of $\mathbf{X}$.

Theorem 1.2.1 *There exists the unique system* $\{\gamma_+(n,k), \gamma_-(n,k) \in M(d; \mathbf{R});$ $0 \leq k < n \leq N\}$ *such that*

$$X(n) = -\sum_{k=0}^{n-1} \gamma_+(n,k)X(k) + \nu_+(n), \tag{1.14}$$

$$X(-n) = -\sum_{k=0}^{n-1} \gamma_-(n,k)X(-k) + \nu_-(-n). \tag{1.15}$$

Here, $\delta_+(n)$, $\delta_-(n)$ $(1 \leq n \leq N)$, which are known as partial correlation functions, and are defined as

$$\delta_+(n) = \gamma_+(n,0),\ \delta_-(n) = \gamma_-(n,0). \tag{1.16}$$

The equations (1.14) and (1.15) are called $\mathrm{KM_2O}$-Langevin equations associated with $\mathbf{X}$. There exist interactions between fluctuation parts and dissipation parts. These interactions are called the Fluctuation–Dissipation Theorem (FDT).
For $n \in \{0, \cdots, N\}$, we set

$$E(\nu_+(n)^t\nu_+(n)) = V_+(n), \text{ and } E(\nu_-(-n)^t\nu_-(-n)) = V_-(n). \tag{1.17}$$

FDT is described as follows:

Theorem 1.2.2 (FDT) *For* $1 \leq k < n \leq N$,

(i) $\gamma_+(n,k) = \gamma_+(n-1,k-1) + \delta_+(n)\gamma_-(n-1,n-1-k)$,

(ii) $\gamma_-(n,k) = \gamma_-(n-1,k-1) + \delta_-(n)\gamma_+(n-1,n-1-k)$,

(iii) $\delta_+(n) = -(R(n) + \sum_{m=0}^{n-2} \gamma_+(n-1,m)R(m+1))V_-(n-1)^{-1}$,

(iv) $\delta_-(n) = -({}^tR(n) + \sum_{m=0}^{n-2} \gamma_-(n-1,m){}^tR(m+1))V_+(n-1)^{-1}$.

For $1 \leq n \leq N$,

(v) $V_+(n) = (I - \delta_+(n)\delta_-(n))V_+(n-1)$,

(vi) $V_-(n) = (I - \delta_-(n)\delta_+(n))V_-(n-1)$.

For the special case of $n = 1$, *we get*

(vii) $V_+(0) = V_-(0) = R(0)$,

(viii) $\delta_+(1) = -R(1)R(0)^{-1}$,

(ix) $\delta_-(1) = -{}^tR(1)R(0)^{-1}$.

When $d = 1$, $R(n) = {}^tR(-n)$. *Therefore, we can see that*

$$\begin{cases} \delta_+(*) = \delta_-(*), \\ \gamma_+(*,\cdot) = \gamma_-(*,\cdot), \\ V_+(*) = V_-(*). \end{cases}$$

FDT is known as the Levinson-Durbin algorithm.
The system $\{\gamma_+(n,k), \gamma_-(n,k), V_+(l), V_-(l); 0 \leq k < n \leq N,\ 0 \leq l \leq N\}$ is called a KM_2O-Langevin data associated with covariance function R.

1.3 Test(S)

Test(S) was proposed by Okabe and Nakano [1991] to test whether given time series are a realization of a local and weakly stationary process or not. Following Nakano [1991], we summarize the deduced processes of Test(S). Any $d, N \in \mathbf{N}$ be fixed. We are given any $N+1$ vectors $\mathcal{D}(n) \in \mathbf{R}^d$ $(0 \leq n \leq N)$. $\mathcal{D} = (\mathcal{D}(n); 0 \leq n \leq N)$ is called data.
The sample mean vector $\mu^{\mathcal{D}}$ of $\mathcal{D}$ and the sample covariance matrix function $R^{\mathcal{D}} = (R^{\mathcal{D}}_{jk})_{1\leq j,k\leq d}$ of $\mathcal{D}$ are defined as follows:

$$\mu^{\mathcal{D}} \equiv \frac{1}{N+1}\sum_{m=0}^{N} \mathcal{D}(m), \tag{1.18}$$

$$R^{\mathcal{D}}_{jk}(n) \equiv \frac{1}{N+1}\sum_{m=0}^{N-n} (\mathcal{D}_j(n+m) - \mu^{\mathcal{D}}_j)(\mathcal{D}_k(m) - \mu^{\mathcal{D}}_k), \tag{1.19}$$

$$R^{\mathcal{D}}_{jk}(-n) \equiv R^{\mathcal{D}}_{kj}(n), \tag{1.20}$$

where

$$\mu^{\mathcal{D}} = \begin{pmatrix} \mu_1^{\mathcal{D}} \\ \vdots \\ \mu_d^{\mathcal{D}} \end{pmatrix}, \ \mathcal{D}(n) = \begin{pmatrix} \mathcal{D}_1(n) \\ \vdots \\ \mathcal{D}_d(n) \end{pmatrix} \qquad (0 \le n \le N). \tag{1.21}$$

The standardized data $\mathcal{X} = (\mathcal{X}(n); 0 \le n \le N)$ of $\mathcal{D}$ is defined as follows:

$$\mathcal{X}(n) = \begin{pmatrix} \sqrt{R_{11}^{\mathcal{D}}(0)^{-1}} & & \mathbf{0} \\ & \ddots & \\ \mathbf{0} & & \sqrt{R_{dd}^{\mathcal{D}}(0)^{-1}} \end{pmatrix} (\mathcal{D}(n) - \mu^{\mathcal{D}}). \tag{1.22}$$

Let $R^{\mathcal{X}} = (R_{jk}^{\mathcal{X}})_{1 \le j,k \le d}$ be the sample covariance matrix function of $\mathcal{X}$ defined similar to (1.18), (1.19), and (1.20).
We can define the sample block Toeplitz matrix $S_n^{\mathcal{X}}$ $(1 \le n \le N)$. Here, it is assumed that

$$S_n^{\mathcal{X}} \in GL(nd; \mathbf{R}) \qquad (1 \le n \le N). \tag{1.23}$$

Replacing R by $R^{\mathcal{X}}$ in the algorithm from (i) to (ix) in Section 1.2, we get the sample KM$_2$O-Langevin data $\{\gamma_+(n,k),\ \gamma_-(n,k),\ V_+(l),\ V_-(l);\ 0 \le k < n \le N,\ 0 \le l \le N\}$.
Then $\nu_+ = (\nu_+(n); 0 \le n \le N)$, which is called the sample random force of data $\mathcal{X}$, is introduced by

$$(*) \begin{cases} \nu_+(0) = \mathcal{X}(0), \\ \nu_+(n) = \mathcal{X}(n) + \sum_{k=0}^{n-1} \gamma_+(n,k)\mathcal{X}(k) \qquad (1 \le n \le N). \end{cases}$$

We choose lower triangular matrices $\Gamma_+(n) \in GL(d; \mathbf{R})$ such that

$$V_+(n) = \Gamma_+(n)^t \Gamma_+(n) \qquad (0 \le n \le N). \tag{1.24}$$

We define the d-dimensional data $\xi_+ = (\xi_+(n); 0 \le n \le N)$ by

$$\xi_+(n) = \Gamma_+(n)^{-1} \nu_+(n) \qquad (0 \le n \le N). \tag{1.25}$$

Set

$$\xi_+(n) = \begin{pmatrix} \xi_{+1}(n) \\ \vdots \\ \xi_{+d}(n) \end{pmatrix} \qquad (0 \le n \le N). \tag{1.26}$$

Rearranging (1.26), we can construct the one-dimensional data $\xi = (\xi(n); 0 \le n \le d(N+1) - 1)$ as follows: For $n = 0, \ldots, d(N+1) - 1$,

$$\xi(n) = \xi_{+p}(m),\ n = dm + p - 1 \qquad (1 \le p \le d,\ 0 \le m \le N). \tag{1.27}$$

Then, the Construction Theorem of Okabe [1998] brings that (S.1) and (S.2) below are equivalent to each other.

(S.1) *$\mathcal{X}$ is a realization of a local and weakly stationary time series with $R^{\mathcal{X}}$ as its covariance function.*

(S.2) *ξ realizes a one-dimensional standardized white noise.*

To test (S.2), we introduce μ^{ξ}, $(v^{\cdot} - 1)^{\sim}$, and $R^{\xi}(n, m)$ $(1 \leq n \leq L_N,\ 0 \leq m \leq L_N - n)$ by

$$\mu^{\xi} = \frac{1}{d(N+1)} \sum_{k=0}^{d(N+1)-1} \xi(k), \tag{1.28}$$

$$(v^{\xi} - 1)^{\sim} = \frac{1}{d(N+1)} \Big(\sum_{k=0}^{d(N+1)-1} \xi(k)^2 \Big), \tag{1.29}$$

$$\times \Big(\sum_{k=0}^{d(N+1)-1} (\xi(k)^2 - 1)^2 \Big)^{-1/2}, \tag{1.30}$$

$$R^{\xi}(n, m) = \frac{1}{d(N+1)} \sum_{k=m}^{d(N+1)-1-n} \xi(k)\xi(n+k). \tag{1.31}$$

Here, L_N is an effective length of R^{ξ} and in this case is taken to be

$$L_N = [2\sqrt{d(N+1)}] - 1. \tag{1.32}$$

We institute the following criterion (M), (V), (O) for checking whether ξ satisfies (S.2) or not.

(M) $\sqrt{d(N+1)}|\mu^{\xi}| < 1.96$,

(V) $|(v^{\cdot} - 1)^{\sim}| < 2.2414$,

(O) for any n, m $(1 \leq n \leq L_N,\ 0 \leq m \leq L_N - n)$.

$$d(N+1)\Big(\sum_{j=1}^{2} (L^{(j)}_{n,m})^{1/2}\Big)^{-1} |R^{\xi}(n, m)| < 1.96. \tag{1.33}$$

Here $L^{(j)}_{n,m}$ $(1 \leq j \leq 2)$ is defined as follows: Dividing $d(N+1)$ and m by $2n$ and n, respectively, we get the following expression form.

$$d(N+1) = q(2n) + r \quad (0 \leq r \leq 2n - 1), \tag{1.34}$$

$$m = sn + t \qquad (0 \leq t \leq n - 1). \tag{1.35}$$

If $r \in \{0, \ldots, n\}$, then

$$L^{(1)}_{n,m} = \begin{cases} n(q + (s/2)) - m & (s \text{ is even}), \\ n(q - (s+1)/2) & (s \text{ is odd}), \end{cases}$$

$$L^{(2)}_{n,m} = \begin{cases} n(q-1-(s/2))+r & (s \text{ is even}), \\ n(q-1+(s+1)/2)+r-m & (s \text{ is odd}). \end{cases}$$

and if $r \in \{n+1, \ldots, 2n-1\}$,

$$L^{(1)}_{n,m} = \begin{cases} n(q-1+(s/2))+r-m & (s \text{ is even}), \\ n(q-1-(s+1)/2)+r & (s \text{ is odd}), \end{cases}$$

$$L^{(2)}_{n,m} = \begin{cases} n(q-(s/2)) & (s \text{ is even}), \\ n(q+(s+1)/2)-m & (s \text{ is odd}). \end{cases}$$

Now, a rule of experience concerning data analysis tells us that an effective number of the sample covariance matrix function $R^{\mathcal{X}}$ is considered to be at most $[3\sqrt{N+1}/d]$. We set

$$M = [3\sqrt{N+1}/d] - 1. \tag{1.36}$$

Making use of the reliable $\{R(n); 0 \le n \le M\}$ and the reliable subsystem $\{\gamma_+(n,k),\ \gamma_-(n,k),\ V_+(l), V_-(l); 0 \le k < n \le M,\ 0 \le l \le M\}$, we restate the new criterion alternative to (M), (V), and (O).
For each $i \in \{0, \ldots, N-M\}$, we consider the shifted data $\mathcal{X}_i$ with $\mathcal{X}(i)$ as its initial point $\mathcal{X}_i(0)$:

$$\mathcal{X}_i = (\mathcal{X}(i+n); 0 \le n \le M). \tag{1.37}$$

Similar to $(*)$, the sample random force $\nu_{+i} = (\nu_{+i}(n);\ 0 \le n \le M)$ of data $\mathcal{X}_i$ is defined by

$$(*)_i \begin{cases} \nu_{+i}(0) = \mathcal{X}(i), \\ \nu_{+i}(n) = \mathcal{X}(i+n) + \sum_{k=0}^{n-1} \gamma_+(n,k)\mathcal{X}(i+k) \qquad (1 \le n \le M). \end{cases}$$

In (1.26) and (1.27) replacing $\xi(n)$ by $\xi_i(n)$, $\xi_{+j}(n)$ $(1 \le j \le d)$ by $\xi_{+ij}(n)$ $(1 \le j \le d)$, and N by M, respectively, the one-dimensional data $\xi_i = (\xi_i(n); 0 \le n \le d(M+1)-1)$ is constructed similarly.
Further, we replace $\xi(n)$ by $\xi_i(n)$ and N by M from (1.28) to (1.31). Then we get the criterion $(\mathrm{M})_i$, $(\mathrm{V})_i$, and $(\mathrm{O})_i$ which check that ξ_i is a realization of a normalized white noise.
Concerning the main problem of testing the local and weak stationarity of the original data $\mathcal{D}$, Okabe and Nakano [1955] proposed:
Test(S): *the rate of* $i \in \{0, \ldots, N-M\}$ *for which* $(\mathrm{M})_i$ *(resp.* $(\mathrm{V})_i$ *and* $(\mathrm{O})_i$*) holds is over 80 percent (resp. 70 percent and 80 percent).*
We say that data $\mathcal{D}$ is a realization of a local and weakly stationary process if Test(S) is accepted. Also, we say simply that $\mathcal{D}$ has the local and weak stationarity.

Chapter 2

Causality Test in Time Series

by Yuji Nakano

We introduce the local causal value, which measures the degree of influence of time series Y on time series X. When X and Y are stationary processes, we propose the Local Causality Test which tests the existence of local causality between X and Y. If a stationary time series Z influences X and Y, we can construct $\epsilon_{X,Z}$ and $\epsilon_{Y,Z}$ eliminating the influence of Z, respectively. Applying the Local Causal Test, we propose the Partial Local Causality Test, which tests that Z is a hidden factor of causality between X and Y.

2.1 Introduction

Up to this day, many works on causal analysis in time series are known [Granger, 1969]. Many studies in time series analysis assume "weak stationarity" of given data and use simplified models such as autoregressive (AR) or autoregressive moving average (ARMA) models fitting for given data.
We apply Test(S) to given data, including some transformed data. If the data pass Test(S), viz. that they are accepted to have stationarity, we proceed to further analysis.
Nakano [1999] proposed a concept of causality, which we call local causality from the viewpoint of the theory of KM_2O-Langevin equations.
The first purpose of this chapter is to introduce the causal value between two stationary time series with finite time space and propose a test which we call the Local Causal Test. The second purpose is as follows. Let time series $\mathcal{X}$, $\mathcal{Y}$ and $\mathcal{Z}$ be given. We assume that $\mathcal{Y}$ gives a causality to $\mathcal{X}$. However, there is a possibility that $\mathcal{Z}$ is one of the important factors of this causal relation.

DOI:10.1201/9781003603429-2

At first, we construct two time series which are eliminated the influence of $\mathcal{Z}$ from $\mathcal{X}$ and $\mathcal{Y}$, respectively.
Secondly, we analyze the causal relation between these time series. We call this process the partial local causality analysis. The Local Causality Test and the Partial Local Causality Test is proposed.

2.2 Local Causality

Let $N \in \mathbb{N}$. $\mathbf{X}=(X(n); 0 \leq n \leq N)$ and $\mathbf{Y}=(Y(n); 0 \leq n \leq N)$ are one-dimensional stochastic processes with mean zero on a probability space $(\Omega, \mathcal{B}, \mathcal{P})$, respectively.
We assume

$$X(n), Y(n) \in L^2(\Omega, \mathcal{B}, \mathcal{P}),\ 0 \leq n \leq N. \tag{2.1}$$

Further, we introduce two-dimensional stochastic process $\mathbf{W}_1=(W_1(n); 0 \leq n \leq N)$ by

$$W_1(n) = \begin{pmatrix} X(n) \\ Y(n) \end{pmatrix}. \tag{2.2}$$

Let J be an information set and $\hat{X}_J(n)$ be the linear prediction $X(n)$ by J. Then, the prediction error of $X(n)$ by J and its variance are defined:

$$\epsilon(X(n)|J) = X(n) - \hat{X}_J(n), \tag{2.3}$$

$$\sigma^2(X(n)|J) = \|\epsilon(X(n)|J)\|^2 =< \epsilon(X(n)|J), \epsilon(X(n)|J) > . \tag{2.4}$$

For each $n \in \{1, \ldots, N\}$, we consider the cases
$J = \mathbf{M}_0^{n-1}(\mathbf{X})$ and
$J = \mathbf{M}_0^{n-1}(\mathbf{X}, \mathbf{Y})$, respectively. Now, local causality between $\mathbf{X}$ and $\mathbf{Y}$ is defined as follows:

Definition 2.2.1 (Local causality) If there exits $n \in \{1, \ldots, N\}$ such that

$$\sigma(X(n)|\mathbf{M}_0^{n-1}(\mathbf{X}, \mathbf{Y})) < \sigma(X(n)|\mathbf{M}_0^{n-1}(\mathbf{X})), \tag{2.5}$$

we say that $\mathbf{Y}$ causes $\mathbf{X}$ locally, denoted by $\mathbf{Y} \stackrel{\text{LC}}{\Longrightarrow} \mathbf{X}$.
Otherwise, we say that $\mathbf{Y}$ does not cause $\mathbf{X}$ locally, denoted by $\mathbf{Y} \stackrel{\text{LC}}{\not\Longrightarrow} \mathbf{X}$.

Definition 2.2.2 Morita defined the local causality value from $\mathbf{Y}$ to $\mathbf{X}$ by

$$\text{LC}(\mathbf{X}, \mathbf{Y}) = \frac{1}{N}\sum_{n=1}^{N}\{\sigma^2(X(n)|\mathbf{M}_0^{n-1}(\mathbf{X})) - \sigma^2(X(n)|\mathbf{M}_0^{n-1}(\mathbf{X}, \mathbf{Y}))\}. \tag{2.6}$$

Theorem 2.2.1 *1.* $\mathbf{Y} \overset{\mathrm{LC}}{\Longrightarrow} \mathbf{X}$ *if and only if* $\mathrm{LC}(\mathbf{X}, \mathbf{Y}) > 0$,

2. $\mathbf{Y} \overset{\mathrm{LC}}{\not\Longrightarrow} \mathbf{X}$ *if and only if* $\mathrm{LC}(\mathbf{X}, \mathbf{Y}) = 0$.

When $\mathbf{X}$ and $\mathbf{W}_1$ have stationarity, we set $\mathrm{KM_2O}$-Langevin data $\mathcal{LM}^+(\mathbf{X})$ and $\mathcal{LM}^+(\mathbf{W}_1)$ by

$$\begin{aligned}\mathcal{LM}^+(\mathbf{X}) &= \{\gamma_{+X}(n,k), V_{+X}(m); \\ &\qquad 0 \le k < n \le N,\ 0 \le m \le N\}, \qquad (2.7)\\ \mathcal{LM}^+(\mathbf{W}_1) &= \{\gamma_{+W_1}(n,k), V_{+W_1}(m); \\ &\qquad 0 \le k < n \le N,\ 0 \le m \le N\}. \qquad (2.8)\end{aligned}$$

For each $m (0 \le m \le N)$, let $V_{+W_1,11}(m)$ be the $(1,1)$-component of $V_{+W_1}(m)$. We have the followig Theorem.

Theorem 2.2.2

$$\mathrm{LC}(\mathbf{X}, \mathbf{Y}) = \frac{1}{N}\sum_{n=1}^{N}(V_{+X}(n) - V_{+W_1,11}(n)). \qquad (2.9)$$

2.2.1 Local Causality Test

Let $\mathcal{X} = (\mathcal{X}(n); 0 \le n \le N)$, $\mathcal{Y} = (\mathcal{Y}(n); 0 \le n \le N)$ be one-dimensional data. $\tilde{\mathcal{X}} = (\tilde{\mathcal{X}}(n); 0 \le n \le N)$, $\tilde{\mathcal{Y}} = \tilde{\mathcal{Y}}(n); 0 \le n \le N)$ are standardization of these data, respectively.
We set $\tilde{\mathcal{W}_1} = ({}^t(\tilde{\mathcal{X}}(n), \tilde{\mathcal{Y}}(n); 0 \le n \le N)$.
We assume that $\tilde{\mathcal{X}}$ and $\tilde{\mathcal{W}_1}$ pass Test(S). It is noted that reliable numbers of the sample covariance functions $R^{\tilde{\mathcal{X}}}$ and $R^{\tilde{\mathcal{W}_1}}$ are at most $1 + M_1 \equiv [3\sqrt{N+1}]$ and $1 + M_2 \equiv \left[\frac{3\sqrt{N+1}}{2}\right]$.
Making use of the reliable $\{V_{+\mathcal{X}}(n), V_{+\mathcal{W}_1}(n); 0 \le n \le M_2\}$, the sample local causal value from $\mathcal{Y}$ to $\mathcal{X}$ is defined by

$$\mathrm{LC}(\tilde{\mathcal{X}}, \tilde{\mathcal{Y}}) = \frac{1}{M_2}\sum_{n=1}^{M_2}(V_{+\tilde{\mathcal{X}}}(n) - V_{+\tilde{\mathcal{W}_1},11}(n)). \qquad (2.10)$$

We introduce one thousand sets of random uniform numbers
$\mathcal{U}_i = (\mathcal{U}_i(n); 0 \le n \le N), i = 1, \ldots, 1000$. $\tilde{\mathcal{U}_i}$ is the standardization of $\mathcal{U}_i$ $(i = 1, \ldots, 1000)$. We get $\mathrm{LC}(\tilde{\mathcal{X}}, \tilde{\mathcal{U}_i}), i = 1, \ldots, 1000$.
Let γ_1 be 90 percentile of data $\mathrm{LC}(\tilde{\mathcal{X}}, \tilde{\mathcal{U}_i}), i = 1, \ldots, 1000$. Namely, we have

$$\frac{1}{1000}\sharp\{i; \mathrm{LC}(\tilde{\mathcal{X}}, \tilde{\mathcal{U}_i}) \le \gamma_1, i = 1, \ldots, 1000\} = 0.90. \qquad (2.11)$$

Now, we propose the criterion of the Local Causal Test.

Definition 2.2.3 (**the Local Causality Test**)

(LC-1) if $\mathrm{LC}(\tilde{\mathcal{X}}, \tilde{\mathcal{Y}}) \leq \gamma_1$, we say $\mathcal{Y} \overset{\mathrm{LC}}{\not\Longrightarrow} \mathcal{X}$.

(LC-2) if $\mathrm{LC}(\tilde{\mathcal{X}}, \tilde{\mathcal{Y}}) > \gamma_1$, we say $\mathcal{Y} \overset{\mathrm{LC}}{\Longrightarrow} \mathcal{X}$.

2.3 Partial Local Causality

We introduce the concept of partial local causality. Let $\mathbf{Z}$=$(Z(n); |n| \leq N)$ be a one-dimensional stationary stochastic process with mean zero on a probability space $(\Omega, \mathcal{B}, \mathcal{P})$.
Two-dimensional stationary stochastic process $\mathbf{W_2} = (W_2(n); |n| \leq N)$@ and $\mathbf{W_3} = (W_3(n); |n| \leq N)$ are defined as

$$W_2(n) = \begin{pmatrix} X(n) \\ Z(n) \end{pmatrix}, \tag{2.12}$$

$$W_3(n) = \begin{pmatrix} Y(n) \\ Z(n) \end{pmatrix}. \tag{2.13}$$

Now, we define the following equations:

$$X(n) = P_{\mathbf{M}_0^n(\mathbf{Z})} X(n) + \epsilon_{(X \backslash Z)}(n), \tag{2.14}$$

$$Y(n) = P_{\mathbf{M}_0^n(\mathbf{Z})} Y(n) + \epsilon_{(Y \backslash Z)}(n). \tag{2.15}$$

We call $\epsilon_{(\mathbf{X} \backslash \mathbf{Z})} = (\epsilon_{(X \backslash Z)}(n); 0 \leq n \leq N)$ and $\epsilon_{(\mathbf{Y} \backslash \mathbf{Z})} = (\epsilon_{(Y \backslash Z)}(n); 0 \leq n \leq N)$ are stochastic processes after eliminating the effect of $\mathbf{Z}$.

Definition 2.3.1 (Partial local causality value) Partial local causality value from $\mathbf{Y}$ to $\mathbf{X}$ after eliminated the effect of $\mathbf{Z}$ is defined by

$$\mathrm{PLC}((\mathbf{X}, \mathbf{Y}) \setminus \mathbf{Z}) = \mathrm{LC}(\epsilon_{(\mathbf{X} \backslash \mathbf{Z})}, \epsilon_{(\mathbf{Y} \backslash \mathbf{Z})}). \tag{2.16}$$

Definition 2.3.2 $\mathbf{Z}$ is said to be the substantial local causality factor which $\mathbf{Y}$ causes $\mathbf{X}$ if

$$\mathrm{LC}(\mathbf{X}, \mathbf{Y})) > 0 \text{ and } \mathrm{PLC}((\mathbf{X}, \mathbf{Y}) \setminus \mathbf{Z}) = 0. \tag{2.17}$$

We have the following $\mathrm{KM_2O}$-Langevin equation of Z.

$$\begin{aligned} Z(n) &= P_{\mathbf{M}_0^{n-1}(\mathbf{Z})} Z(n) + \nu_{+Z}(n) \\ &= -\sum_{i=0}^{n-1} \gamma_{+Z}(n, i) Z(i) + \nu_{+Z}(n). \end{aligned} \tag{2.18}$$

We set

$$P_{\mathbf{M}_0^n(\mathbf{Z})}X(n) = \sum_{k=0}^{n} C_{(X,Z)}(n,k)\nu_{+Z}(k), \tag{2.19}$$

$$P_{\mathbf{M}_0^n(\mathbf{Z})}Y(n) = \sum_{k=0}^{n} D_{(Y,Z)}(n,k)\nu_{+Z}(k). \tag{2.20}$$

Okabe and Inoue [1994] gives

$$C_{(X,Z)}(n,k) = \begin{cases} R_{12}^{\mathbf{W}_2}(n)/R^{\mathbf{Z}}(0) & (k=0), \\ \{R_{12}^{\mathbf{W}_2}(n-k) + & \sum_{i=1}^{k-1}\gamma_{+Z}^{0}(k,i)R_{12}^{\mathbf{W}_2}(n-i)\}/V_{+Z}(k) \\ & (k=1,\ldots,n). \end{cases} \tag{2.21}$$

Here, $R_{12}^{\mathbf{W}_2}$ is the (1,2)-component of $R^{\mathbf{W}_2}$. Namely, we have $R_{12}^{\mathbf{W}_2}(n-k) =< X(n), Z(k) >$ and $R^{\mathbf{Z}}(0) =< Z(0), Z(0) >$. Similarly, we get $D_{(Y,Z)}(n,k)(k = 0,\ldots,n)$. For each n, we express simply

$$\epsilon_X(n) = \epsilon_{(X\setminus Z)}(n), \ \ \epsilon_Y(n) = \epsilon_{(Y\setminus Z)}(n). \tag{2.22}$$

2.3.1 Covariance of Fundamental Variables

For $n, m \in \{0,\ldots,N\}$, $< \epsilon_X(n), \epsilon_X(m) >$ and $< \epsilon_X(n), \epsilon_Y(m) >$ are constructed by Langevin data of Z and covariance function of W_2. We show the case of $< \epsilon_X(n), \epsilon_Y(m) >$. Let $s = \min\{n,m\}$.

$$\begin{aligned} &< \epsilon_X(n), \epsilon_Y(m) > =< X(n), Y(m) > \\ &- \sum_{l=0}^{m} < X(n), D_{(Y,Z)}(m,l)\nu_{+Z}(l) > - \sum_{k=0}^{n} < C_{(X,Z)}(n,k)\nu_{+Z}(k), Y(m) > \\ &+ \sum_{k=0}^{n}\sum_{l=0}^{m} C_{(X,Z)}(n,k)D_{(Y,Z)}(m,l) < \nu_{+Z}(k), \nu_{+Z}(l) > \\ &= R^{X,Y}(n-m) - \sum_{k=0}^{m} D_{(Y,Z)}(m,l) < X(n), \nu_{+Z}(l) > \\ &- \sum_{k=0}^{n} C_{(X,Z)}(n,k) < \nu_{+Z}(k), Y(m) > + \sum_{k=0}^{s} C_{(X,Z)}(n,k)D_{(Y,Z)}(m,k)V_Z(k). \end{aligned} \tag{2.23}$$

Here,

$$\begin{aligned} < X(n), \nu_{+Z}(l) > &=< X(n), Z(l) > + \sum_{i=0}^{l-1} < X(n), \gamma_{+Z}(l,i) Z(i) > \\ &= R^{X,Z}(n-l) + \sum_{i=0}^{l-1} \gamma_{+Z}(l,i) R^{X,Z}(n-i). \end{aligned} \quad (2.24)$$

$$\begin{aligned} < \nu_{+Z}(k), Y(m) > &=< Z(k), Y(m) > + \sum_{i=0}^{k-1} < \gamma_{+Z}(k,i) Z(i), Y(m) > \\ &= R^{Z,Y}(k-m) + \sum_{i=0}^{k-1} \gamma_{+Z}(k,i) R^{Z,Y}(i-m). \end{aligned} \quad (2.25)$$

We get
$< \epsilon_X(n), \epsilon_X(m) >$,
$< \epsilon_Y(n), \epsilon_X(m) >$ and $< \epsilon_Y(n), \epsilon_Y(m) >$
in a similar way.

2.3.2 Prediction Error 1

We define orthogonal process $\eta_X(0), \ldots \eta_X(N)$ by

$$\eta_X(0) = \epsilon_X(0), \quad (2.26)$$
$$\eta_X(n) = \epsilon_X(n) - P_{\mathbf{M}_0^{n-1}(\epsilon_X)} \epsilon_X(n) \ (1 \leqq n \leqq N). \quad (2.27)$$

We put

$$p(n,i) =< \epsilon_X(n)), \epsilon_X(i) > \quad (0 \leqq i \leqq n), \quad (2.28)$$
$$q(n,i) =< \epsilon_X(n)), \eta_X(i) > \quad (0 \leqq i \leqq n), \quad (2.29)$$
$$q_\eta(n) = q(n,n) =< \eta_X(n)), \eta_X(n) > . \quad (2.30)$$

Let $\omega_n^X = \{p(i,j)); 0 \leqq i, j \leqq n\}$. We show $q_\eta(n)$ which we call the prediction error I is constructed by ω_n^X.
If $q(n,i)$ is constructed by ω_n^X through the steps, we say $q(n,i)$ is calculated.
At first, we have

$$q_\eta(0) =< \eta_X(0)), \eta_X(0) >=< \epsilon_X(0)), \epsilon_X(0) >= p(0,0). \quad (2.31)$$

Let $n = 1$.

$$\begin{aligned} \eta_X(1) &= \epsilon_X(1) - a_{1,0}\eta_X(0) = \epsilon_X(1) - a_{1,0}\epsilon_X(0), \\ a_{1,0} &= \frac{< \epsilon_X(1), \epsilon_X(0) >}{< \epsilon_X(0), \epsilon_X(0) >} = \frac{p(1,0)}{p(0,0)}. \end{aligned} \quad (2.32)$$

$$\begin{aligned} q_\eta(1) &=< \eta_X(1), \eta_X(1) >=< \epsilon_X(1), \eta_X(1) > \\ &=< \epsilon_X(1), \epsilon_X(1) > -a_{1,0} < \epsilon_X(1), \epsilon_X(0) >= p(1,1) - a_{1,0}p(1,0). \end{aligned} \quad (2.33)$$

For $m \geqq 1$.

$$\begin{aligned} q(m,0) &=< \epsilon_X(m), \eta_X(0) >=< \epsilon_X(m), \epsilon_X(0)) >= p(m,0), \\ q(m,1) > &=< \epsilon_X(m), \epsilon_X(1) > -a_{1,0} < \epsilon_X(m), \epsilon_X(0)) > \\ &= p(m,1) - a_{1,0}p(m,0). \end{aligned} \quad (2.34)$$

Therefore, $q(m,0), q(m,1)$ are calculated for any m. Now, we have

$$\eta_X(2) = \epsilon_X(2) - a_{2,0}\eta_X(0) - a_{2,1}\eta_X(1) \quad (2.35)$$

$$a_{2,i} = \frac{< \epsilon_X(2)), \eta_X(i) >}{< \eta_X(i), \eta_X(i) >} = \frac{p(2,i)}{q_\eta(i)} \quad (i = 0,1). \quad (2.36)$$

$$\begin{aligned} q(m,2) &= < \epsilon_X(m), \eta_X(2) > \\ &= < \epsilon_X m), \epsilon_X(2) > - \sum_{i=0}^{1} a_{2,i} < \epsilon_X m), \eta_X(i > \\ &= p(m,2) - \sum_{i=0}^{1} a_{2,i} q(m,i). \end{aligned} \quad (2.37)$$

Therefore, $q(m,2)$ is calculated for any m. Let $k \in \{3 \leqq k \leqq n\}$. Inductively, we get $\eta_X(k)$ and $q_\eta(n)$ as the followings. When $\eta_X(i)$ $(i = 0, ..., k-1)$ are given and $q(m,i)$ $(i = 0, ..., k-1)$ are calculated for any m, we have

$$\begin{aligned} \eta_X(k) &= \epsilon_X(k) - \sum_{l=0}^{k-1} a_{k,l}\eta_X(l), a_{k,l} = \frac{q(k,l)}{q_\eta(l)}. \\ q(m,k) &= p(m,k) - \sum_{l=0}^{k-1} a_{k,l} q(m,l) \ (1 \leqq m \leqq N). \end{aligned} \quad (2.38)$$

This shows $q(m,k)$ is calculated. Now, we have for any n,

$$q_\eta(n) = q(n,n) = p(n,n) - \sum_{l=0}^{n-1} a_{n,l} q(n,l). \quad (2.39)$$

We call $q_\eta(n)$ the prediction error 1 at time n.

2.3.3 Prediction Error 2

T and U are two-dimensional random variables expressed as

$$T = \begin{pmatrix} T_1 \\ T_2 \end{pmatrix}, U = \begin{pmatrix} U_1 \\ U_2 \end{pmatrix}. \quad (2.40)$$

We note

$$[T, U] = E(T^t U) = \begin{pmatrix} < T_1, U_1 > & < T_1, U_2 > \\ < T_2, U_1 > & < T_2, U_2 > \end{pmatrix}. \tag{2.41}$$

Let $n \in \{0, ...N\}$. We set

$$\epsilon(n) = \begin{pmatrix} \epsilon_X(n) \\ \epsilon_Y(n) \end{pmatrix}, \tag{2.42}$$

$$\phi(n) = \begin{pmatrix} \phi_1(n) \\ \phi_2(n) \end{pmatrix}. \tag{2.43}$$

Here,

$$\begin{aligned} \phi(0) &= \epsilon(0), \\ \phi(n) &= \epsilon(n) - P_{\mathbf{M}_0^{n-1}(\mathbf{ff})}\epsilon(n) \ (1 \leqq n \leqq N). \end{aligned} \tag{2.44}$$

We put

$$\begin{aligned} P(n, i) &= [\epsilon(n), \epsilon(i)], Q(n, i) = [\epsilon(n), \phi(i)] \ (i = 0, \ldots, n), \\ Q_\phi(n) &= [\phi(n), \phi(n)]. \end{aligned} \tag{2.45}$$

Let $\Omega_n = \{P(i, j)); 0 \leqq i, j \leqq n\}$. $Q(n, i)$ is calculated by Ω_n through the steps. We find $< \phi_1(n), \phi_1(n) >$ which is the (1,1) component of $[\phi(n), \phi(n)]$. We call $< \phi_1(n), \phi_1(n) >$ the prediction error 2 at time n.
We can find the prediction error 2 similar to the prediction error 1,

$$P_{\mathbf{M}_0^{n-1}(\mathbf{ff})}\epsilon(n) = A_{n,0}\phi(0) + \cdots + A_{n,n-1}\phi(n-1) \quad (n \geqq 1). \tag{2.46}$$

Let $n = 1$.

$$\begin{aligned} \phi(1) &= \epsilon(1) - A_{1,0}\phi(0) = \epsilon(1) - A_{1,0}\epsilon(0), \\ A_{1,0} &= P(1, 0)P(0, 0)^{-1}. \end{aligned} \tag{2.47}$$

$$\begin{aligned} Q(1, 1) &= [\phi(1), \phi(1)] = [\epsilon(1), \phi(1)] \\ &= [\epsilon(1), \epsilon(1) - A_{1,0}\phi(0)] = P(1, 1) - Q(1, 0)^t(A_{1,0}). \end{aligned} \tag{2.48}$$

For any $m \geqq 1$,

$$\begin{aligned} Q(m, 0) &= [\epsilon(m), \phi(0)] = [\epsilon(m), \epsilon(0)] = P(m, 0), \\ Q(m, 1) &= [\epsilon(m), \phi(1)] = P(m, 1) - P(m, 0)^t(A_{1,0}). \end{aligned} \tag{2.49}$$

Inductively, we get $[\phi(n), \phi(n)] \quad (n \geqq 2)$. For $k \in \{2, ...n\}$,

$$\begin{aligned} \phi(k) &= \epsilon(k) - \sum_{l=0}^{k-1} A_{k,l}\phi(l), \\ A_{k,l} &= Q(k, l)Q_\phi(l)^{-1}. \end{aligned} \tag{2.50}$$

$$
\begin{aligned}
Q(n,k) &= [\epsilon(n),\phi(k)] = [\epsilon(n),\epsilon(k) - \sum_{l=0}^{k-1}[A_{k,l}\phi(l)] \\
&= P(n,k) - \sum_{l=0}^{k-1} Q(n,l)^t(A_{k,l}).
\end{aligned} \tag{2.51}
$$

By induction, we have

$$
\begin{aligned}
Q_\phi(n) &= [\epsilon(n),\phi(n)] = [\epsilon(n),\epsilon(n) - \sum_{l=0}^{n-1} A_{n,l}\phi(l)] \\
&= P(n,n) - \sum_{l=0}^{n-1} Q(n,l)^t(A_{n,l}).
\end{aligned} \tag{2.52}
$$

2.3.4 Partial Local Causality Test

Now, we have

$$
\begin{aligned}
\mathrm{PLC}((\mathbf{X},\mathbf{Y}) \setminus \mathbf{Z}) &= \mathrm{LC}(\epsilon_{(\mathbf{X}\setminus\mathbf{Z})},\epsilon_{(\mathbf{Y}\setminus\mathbf{Z})}) \\
&= \frac{1}{N}\sum_{n=1}^{N}(q_\eta(n) - < \phi_1(n),\phi_1(n) >).
\end{aligned} \tag{2.53}
$$

Definition 2.3.3 (**Sample Partial Local Causality Value**)
$\tilde{\mathcal{Z}} = (\mathcal{Z}(n); 0 \leq n \leq N)$ is the standardization of the data $\mathcal{Z} = (\mathcal{Z}(n); 0 \leq n \leq N)$.
Sample local causality value from $\tilde{\mathcal{Y}}$ to $\tilde{\mathcal{X}}$ after eliminating the effect of $\tilde{\mathcal{Z}}$ is defined by

$$
\begin{aligned}
\mathrm{PLC}((\tilde{\mathcal{X}},\tilde{\mathcal{Y}}) \setminus \tilde{\mathcal{Z}}) &= \mathrm{LC}(\epsilon_{(\tilde{\mathcal{X}}\setminus\tilde{\mathcal{Z}})},\epsilon_{(\tilde{\mathcal{Y}}\setminus\tilde{\mathcal{Z}})}) \\
&= \frac{1}{M_2}\sum_{n=1}^{M_2}(< \eta_{\tilde{X}}(n),\eta_{\tilde{X}}(n) > - < \tilde{\phi}_1(n),\tilde{\phi}_1(n) >).
\end{aligned} \tag{2.54}
$$

Here, $\tilde{\phi}_1$ is the correspondent ϕ_1 for $\tilde{\mathcal{X}}$, $\tilde{\mathcal{Y}}$ and $\tilde{\mathcal{Z}}$.
Similar to (2.11), γ_2 is defined by

$$
\frac{1}{1000}\sharp\{i; \mathrm{PLC}((\tilde{\mathcal{X}},\tilde{\mathcal{Y}}) \setminus \tilde{\mathcal{U}}_i) \leq \gamma_2, i = 1,\ldots,1000\} = 0.90. \tag{2.55}
$$

Definition 2.3.4 (**the Partial Local Causality Test**)
If $\mathrm{LC}(\tilde{\mathcal{X}},\tilde{\mathcal{Y}}) > \gamma_1$ and $\mathrm{PLC}((\tilde{\mathcal{X}},\tilde{\mathcal{Y}}) \setminus \tilde{\mathcal{Z}}) \leq \gamma_2$, $\mathcal{Z}$ is called the partial local causality factor which $\mathcal{Y}$ causes $\mathcal{X}$.

Part II

Applications

Chapter 3

Climate Impact of the Mt. Pinatubo Eruption in 1991

by Osamu Morita

In this chapter, we will discuss the effect of the Mt. Pinatubo eruption in 1991 on the global climate. The causal analysis and the correlation analysis are used for this study, and it will be shown that the cooperative effect of the large volcanic eruption and the ENSO warm event is crucial on the global climate, and that the causal analysis is the very useful tool for time sequential data analysis.

3.1 Introduction

The explosive eruption of Mt. Pinatubo (Luzon Island, Philippines) on June 12th, 1991 was the largest in the 20th century, and 20 to 30 megaton sulfur dioxide was injected into the lower stratosphere [Bluth et al, 1992; McCormick and Veiga, 1992]. The magnitude of the eruption was very large and the dust veil index (DVI) was 1000 [Robock and Mao, 1995; Jones and Kelly, 1996]. DVI was defined by Lamb [1970, 1977, 1983] as the measure of the volcanic impact on the climate system. The reference value is 1000 defined for the eruption of Krakatau (Indonesia) in 1883. Only two single eruptions reached the value after the 19th century, i.e. the case of Krakatau and Mt. Pinatubo. Three successive eruptions in 1902 at Latin America reached 1000 by the total; Pelée (Martinique Island, West Indies, DVI = 100), Soufriére (St. Vincent Island, West Indies, DVI = 300), and Santa Maria (Guatemala, DVI = 600) [Lamb, 1983].
Sulfur dioxide injected by Mt. Pinatubo into the lower stratosphere reacted with hydroxyl ion, and formed hydrated sulfuric acid aerosols (SAAs), which

DOI:10.1201/9781003603429-3

stayed in the lower stratosphere for several years. The SAAs reached the maximum in February 1992 and decreased exponentially with the e-folding time of around one year [Robock and Mao, 1995; Fiocco et al., 1996]. The SAAs increased terrestrial albedo (known as the parasol effect), and played the primary role on the cooling of the globe in 1992 and 1993.
Many studies have been made about the volcanic effects on the climate system. They are divided into analyses of the troposphere and lower stratosphere. Many studies concerning the effect of a single volcanic eruption on the lower stratosphere have been made because of the following reasons:

1. Response of the air temperature of the lower stratosphere is so quick that the cause-and-effect relationship is clear.

2. The volcanic effect always appears as positive air temperature anomalies, whose absolute value is very large, e.g. 10°C in the cases of El Chichón and Mt. Pinatubo.

3. Another climate signal comparable to the volcanic signal is only one owing to the quasi-biennial oscillations (QBOs) [Labitzke et al., 1983: Parker and Branscombe, 1983: Angell and Korshover, 1983a, b: Fujita, 1985: Angell, 1996: Labitzke and van Loon, 1996].

On the other hand, the volcanic impact on the climate of the troposphere is very complicated as follows:

1. Volcanic impacts appear after at least half a year and continue a few years following volcanic eruptions.

2. The surface air temperature ($\mathbb{SAT}$) response is not so simple as the $\mathbb{SAT}$ anomalies ($\mathbb{SAT}$As) are generally negative but are positive at high latitudes of the winter hemisphere [Robock and Mao, 1992].

3. There are several climate signals comparable to the volcanic signal, e.g. the signal of the El Niño and Southern Oscillation (ENSO), the North Atlantic Oscillation (NAO), and so on.

Due to the reasons mentioned above, the statistical technique called the superposed epoch analysis (SEA) was employed in many studies [Mitchell, 1992: Kelly and Sear, 1984: Sear et al., 1987: Robock and Mao, 1995: Jones and Kelly, 1996]. The SEA enables to pick up a weak signal embedded in climate noises. Robock and Mao [1995] studied in detail superposed $\mathbb{SAT}$ responses of six major volcanic eruptions (Krakatau, Santa Maria, Katmai, Agung, El Chichón, and Pinatubo), and isolated the pure volcanic forcing on the $\mathbb{SATA}$ by removing the ENSO signal using the linear regression method. Jones and Kelly [1996] pointed out that superposed $\mathbb{SAT}$ deviation of four historical volcanic eruptions in the tropics resembled the response of the Mt. Pinatubo eruption. Their results show that the magnitude of the eruption of Mt. Pinatubo was large enough to obtain the statistically significant result from a single eruption.

Thus, it becomes possible to make clear the cause-and-effect relationship between the SAAs and the SATAs. In section 3.2, we describe the weather extremes in 1993. The data and the data processing, and the result will be followed in sections 3.3 and 3.4. Finally, the discussion and conclusions are presented in section 3.5.

3.2 The Weather Extremes in 1993

The weather extremes occurred at many places in the world in 1993. The heavy rainfall during June and July in North America caused the historical flood of the Mississippi River and St. Louis was flooded all over the city. This catastrophic flood was said to occur once in 500 years. In southeast China, a destructive flood occurred in the downstream region of the Yangtze River, which caused about 1000 fatalities. A long and heavy rainfall caused floods also in southeastern France and northern Italy. In northern Japan, a bad crop was caused by a cool summer and we had no rice harvest there at all. In southwestern Japan, a heavy rainfall occurred in Kagoshima prefecture (the south most prefecture of Kyushu Island), the Kotsuki River which is the largest in the prefecture flooded, and three stone bridges out of five on the river were washed away, which were built in the Edo era about 200 years ago. Figure 3.1 shows the ratio of the annual precipitation of 1993 to the long-term (1991–2020) mean for 153 meteorological observatories in Japan. The largest value is 1.765 of the Kagoshima meteorological observatory (st88317). The annual precipitation of 1993 at the station is 4022.0 mm, which is the maximum value since it was established in 1883. Observatories whose ratio exceed 1.6 distribute along the southern coast of the Japan archipelago, which is caused by the stagnation of a Baiu front (a subtropical front) without progressing towards the north as it does in the usual year. We show monthly precipitation of the Kagoshima meteorological observatory in 1993, 1905, and the long-term (1991–2020) mean in Figure 3.2.

The monthly precipitation of 1993 exceeds largely that of the long-term mean from June to September, which is the proof of the stagnation of the Baiu front. The second maximum of the annual precipitation of the Kagoshima meteorological observatory occurred in 1905, and the monthly precipitation resembles to that of 1993. So, we made the same figure as Figure 3.1 for 1905. In spite of the sparse network of meteorological observatories, stations whose ratio of the 1905 annual precipitation to the long-term (1901–1930) mean exceeds 1.2 concentrate around the southern coast of Japan (Figure 3.3). From this figure, we suspect that the behavior of the Baiu front in 1905 resembled that of 1993. It is interesting that the year 1905 is three years after the successive eruptions of three volcanos in Latin America whose total DVI was 1000. A cool summer following the eruption of the Mt. Pinatubo seems primarily due to the parasol effect of the volcanic aerosols. However, if the SAAs were the only cause of the cool summer, it should be more remarkable

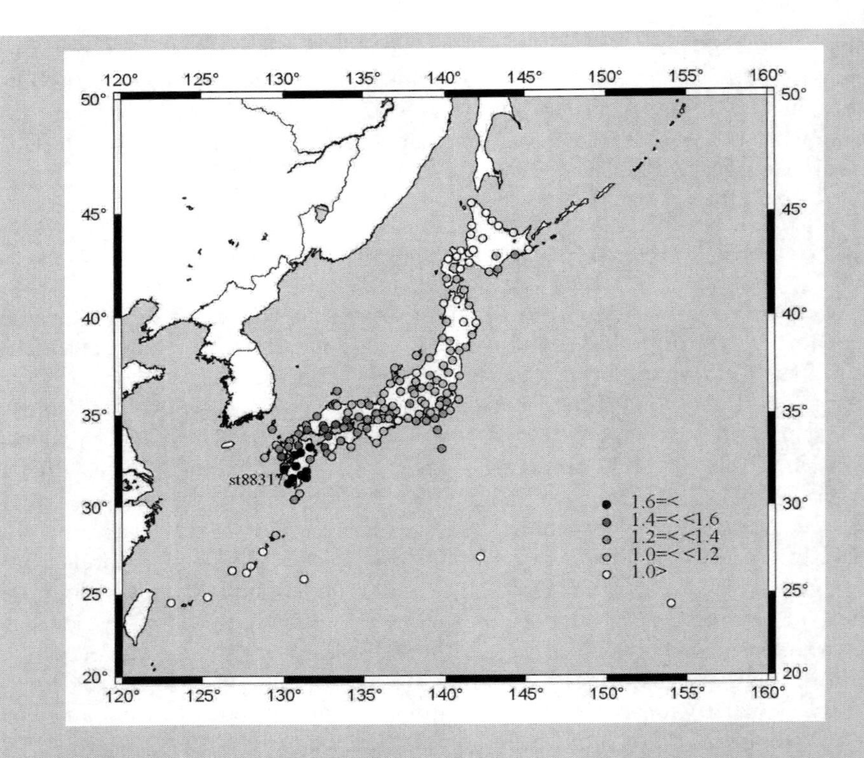

Figure 3.1: The ratio of the annual precipitation of 1993 to the long-term mean (1991–2020) for 153 meteorological observatories in Japan. The station number st88317 denotes the Kagoshima meteorological observatory.

in the summer of 1992 than in the summer of 1993. This fact suggests that the other climate parameters must be taken into account. In fact, a weak El Niño was occurring in 1993, and it is well known that the El Niño warms the tropics and is responsible for a cool summer and a warm winter in Japan. It is the main purpose of this research to make clear the cooperative effect of the SAAs and the El Niño or ENSO warm event on the $\mathbb{SAT}$As.

3.3 Data Used for This Study

Data used for this study are as follows:

1. The monthly air temperature at 1000 hPa level (SAT) of the NCEP/NCAR Reanalysis 1 data [Kalnay et al., 1996][1] from 20°N to 50°N and from 0°E to 360°E (144×13 grid pints).

[1] These data are provided by the Physical Sciences Laboratory (PSL), NOAA, Boulder, Colorado, USA, from their website at https://psl.noaa.gov.

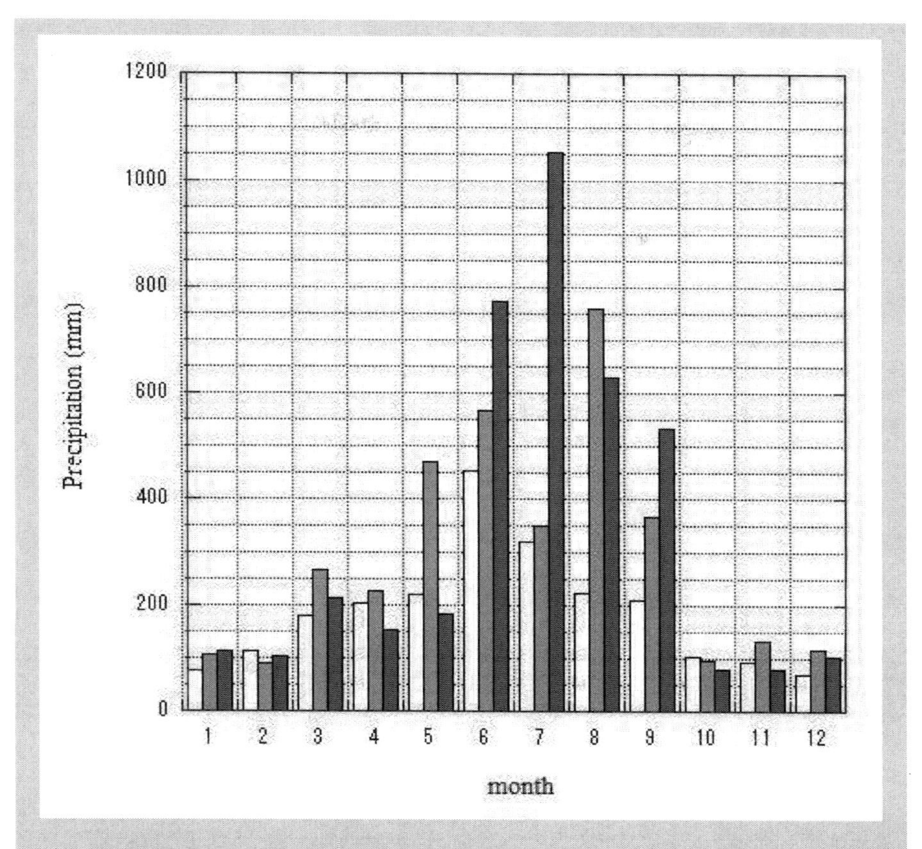

Figure 3.2: The monthly precipitation of the Kagoshima meteorological observatory in 1993 (black bars), 1905 (gray bars), and the long-term (1981–2010) mean (white bars).

2. The monthly amount of SAAs by SAGE II instrument at 3 latitude bands (the 20°N–30°N, the 30°N–40°N, and the 40°N–50°N latitude band). The data period is from January 1985 to December 1996.

3. The monthly surface precipitation (SPR) of Japan provided by Japan Meteorological Agency (JMA). The data period is from the beginning of each meteorological observatory to December 2020.

4. The monthly Southern Oscillation Index (SOI)[2], which is defined by the difference of the normalized sea level pressure (SLP) between Tahiti and Darwin [Ropelewski and Jones, 1987; Allan et al., 1991; Können et al., 1998].

[2]These data are attributed to Climatic Research Unit, University of East Anglia.

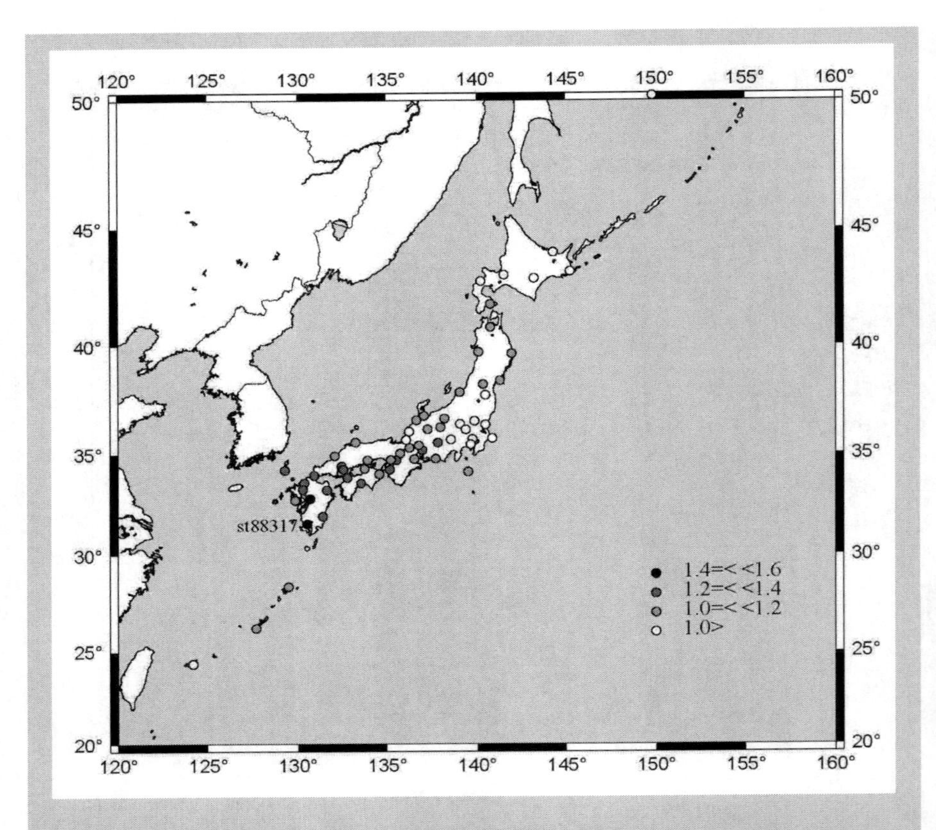

Figure 3.3: The ratio of the annual precipitation of 1905 to the long-term (1901–1930) mean for 62 meteorological observatories in Japan. The station number st88317 denotes the Kagoshima meteorological observatory.

As for the SAT of NCEP/NCAR Reanalysis 1 data, we downloaded the data from January 1981 to December 2010 and calculated the long-term (1981–2010) mean of each month (climatologies) at 144×21 grid points, which were extracted from the actual monthly SAT to make anomalies (SATAs). Thus, we can remove the effect of the seasonal change from the SAT. Finally, we used the SATAs for the data period from January 1985 to December 1996.

Next, we make latitude mean of the SATAs from 20°N to 50°N by every 2.5°, and made 3 latitude band mean SATAs for 20°N–30°N, 30°N–40°N, and 40°N–50°N latitude band, taking into account the weight of the latitude factor.

The time sequence of the SAAs and the SATAs at the 30°N–40°N latitude band is shown in Figure 3.4. We make 7-month moving average of the SATAs. The minimum SATAs occurred at August 1993, and the second minimum

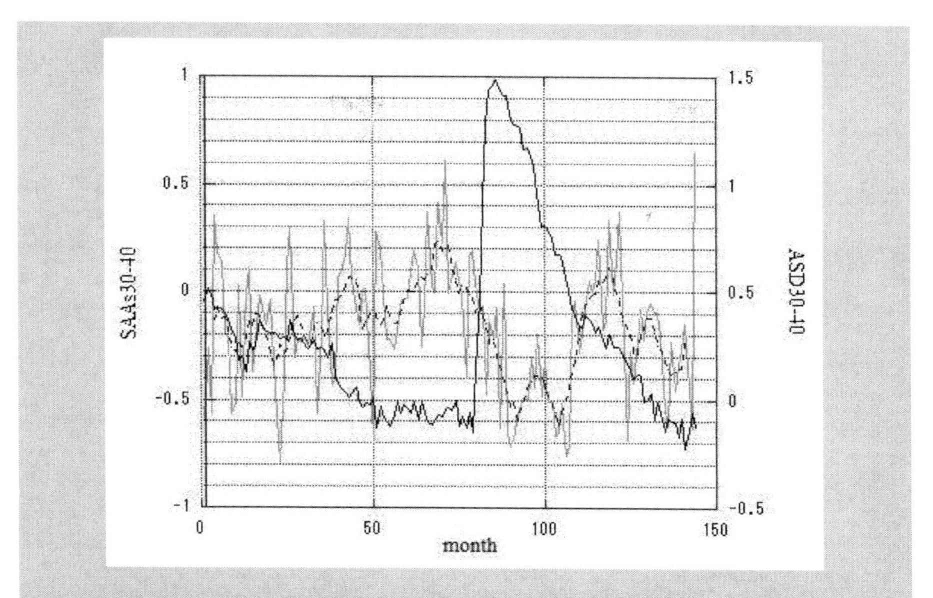

Figure 3.4: Time sequence of the SAAs (a black solid line), the SATAs (a gray solid line), and the 7-month moving average of the SATAs (a black broken line) for the 30°N–40°N latitude band. The unit of the SAAs is $\mu m^2 cm^{-3}$ and that of the SATAs is C°.

occurred at August 1992. At the 20°N–30°N latitude band, the minimum SATAs occurred at December 1992, and the second minimum occurred at February 1993. At the 40°N–50°N latitude band, the minimum SATAs occurred at August 1993, and the second minimum occurred at September 1992.
The SAAs become maximum at February 1992 at all 3 latitude bands and the maximum values are almost the same. These figures show that the sulfur dioxide injected into the tropical lower stratosphere was transported to both poles by Brewer-Dobson circulation reacting with hydroxyl ions and being changed to the SAAs. The SAAs increased almost linearly, reached its maximum at February 1992, and decreased exponentially.
Hence two questions occur:

1. Why the minimum SATAs occurred prior to the maximum SAAs at the lowest latitude band?

2. Why the minimum SATAs occurred in the second summer not in the first summer after the Pinatubo eruption, in spite that the SAAs was around three times larger in the first summer than in the second summer?

These questions will be made clear in the following sections.

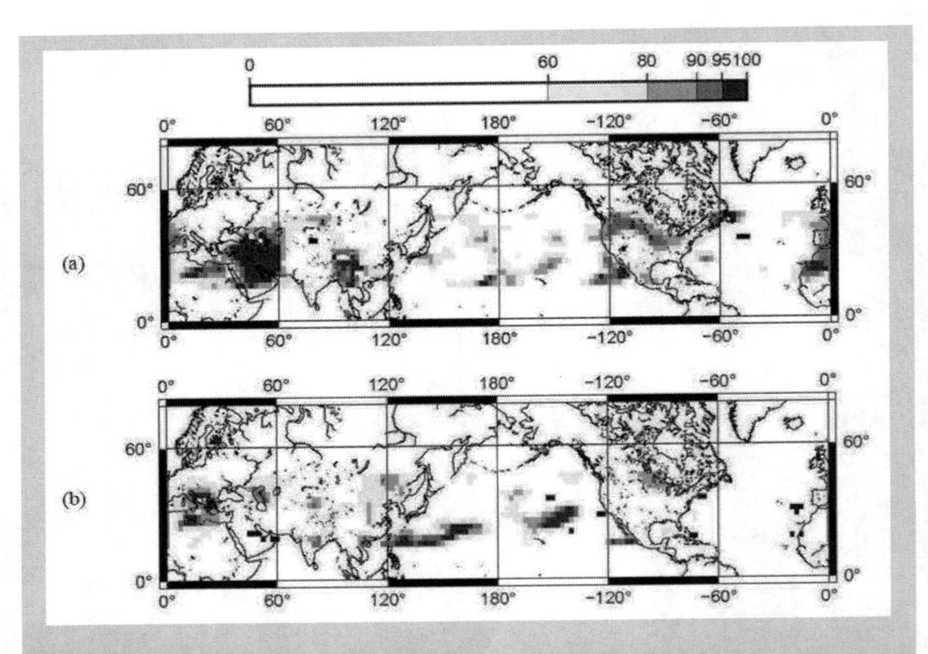

Figure 3.5: The upper panel (a) shows the spatial structure of the percentile of the causal value from the SAAs to the SATAs, and the lower panel (b) is the same as (a) except that the SATAs are removed the influence of the SOI.

3.4 Results of the Analysis

In this section, the causal analysis is applied to the SATAs of 144×13 grid points, 13 latitude circles, and 3 latitude bands. The correlation analysis is also applied to 13 latitude circles, and 3 latitude bands.

3.4.1 Analysis of 144×13 Grid Points

The causal analysis is performed for the SATAs at 144×13 grid points, which locate from the 20° N latitude circle to the 50° N latitude circle. The spatial structure of the percentile of the causal value from the SAAs to the SATAs is shown in Figure 3.5(a). The regions where the SAAs has an intense effect on the SATAs (the percentile of causal value is higher than 95[3]) are northwestern Africa, southern Europe, the Arabian Peninsula, the Middle East, Bangladesh, the western open sea of North America, the east coast of Canada, and the

[3]Nakano [1995] made the criterion that there is a causal influence when the percentile of the causal value is 90–95 and there is a strong causal influence when it is 95–100. The author thinks that this criterion is loose and should be changed to 95–98 and 98–100.

central and eastern tropical Pacific. These regions include the area where weather extremes occurred in 1993 as is described in section 3.2.
Next, the influence of the SOI is removed from the SATAs and the percentile of the causal value from the SAAs to the SATAs removed the SOI is calculated for 144×13 grid points. The spatial structure of the percentile of the causal value is shown in Figure 3.5(b).
The regions where the SAAs has a strong effect on the SATAs removed the SOI are the South China Sea, the central tropical Pacific, the eastern North Pacific Ocean, northwestern Africa, southern Europe, the Arabian Peninsula, the Middle East, Nepal, and northern Vietnam. Comparing Figure 3.5(b) with Figure 3.5(a), we find that the ENSO warm event weakens the cooling effect of the volcanic aerosols in the tropics and intensifies it in the midlatitudes.

3.4.2 Analysis of 13 Latitude Circles

The causal values from the SAAs to the SATAs are calculated for 13 latitude circles, and the percentiles of the causal value are obtained. They are shown in Table 3.1. The maximum negative correlation coefficient and the time lag are also calculated and shown in the table. Causal analysis shows that the SAAs do not affect the SATAs at the lowest latitude circle (the 20.0° N latitude circle) and the 42.5° N latitude circle. The SAAs strongly affect the SATAs from the 25.0° N to the 37.5° N latitude circle and at 47.5° N latitude circle. The SAAs affect the SATAs at the 22.5° N, 40.0° N, 45.0° N, and 50.0° N latitude circle.

Table 3.1: The percentile of the causal value from the SAAs to the SATAs and correlation coefficients between the SAAs and the SATAs at 13 latitude circles.

latitude	correlation coefficient	time lag (month)	percentile of causal value
50.0° N	−0.4700	7	94.5
47.5° N	−0.5066	7	95.3
45.0° N	−0.5034	7	91.1
42.5° N	−0.4897	7	84.1
40.0° N	−0.5354	7	90.3
37.5° N	−0.6078	5	97.8
35.0° N	−0.6564	5	99.8
32.5° N	−0.6412	5	99.8
30.0° N	−0.6200	2	100.0
27.5° N	−0.6450	1	100.0
25.0° N	−0.6273	−2	99.8
22.5° N	−0.6646	−2	90.4
20.0° N	−0.5234	−2	82.1

Table 3.2: The percentile of the causal value from the SAAs to the SATAs removed the SOI, and the correlation coefficients between the SAAs and the SATAs removed the SOI at 13 latitude circles.

latitude	correlation coefficient	time lag (month)	percentile of causal value
50.0° N	−0.4021	7	86.4
47.5° N	−0.4258	7	82.3
45.0° N	−0.4198	7	75.9
42.5° N	−0.4056	7	62.5
40.0° N	−0.4492	7	70.3
37.5° N	−0.5143	6	83.8
35.0° N	−0.5702	5	93.1
32.5° N	−0.5648	5	96.1
30.0° N	−0.5511	5	98.0
27.5° N	−0.5912	1	99.1
25.0° N	−0.6064	1	98.3
22.5° N	−0.5945	1	95.9
20.0° N	−0.5872	1	95.9

The correlation analysis shows the consistent tendency, namely the correlation coefficients between the SAAs and the SATAs are statistically significant over 98% confidential level at all 13 latitude circles. The absolute value of the correlation coefficients is over 0.6 from the 22.5° N to the 37.5° N latitude circle. The time lag n, at which the maximum negative correlation coefficient occurs, means that the negative maximum of the SATAs lag n month from that of the SAAs. The time lag −2 from the 20.0° N to the 25.0° N latitude circle is very curious, because it is unreasonable that the maximum of the SAAs lags that of the SATAs. While, the time lags are positive at other latitude circles and they increase along with latitude.

Next, we remove the effect of the ENSO from the SATAs using the theory of the partial local causality (refer Chapter 2). The causal values from the SAAs to the SATAs removed the effect of the SOI are calculated and the percentile of the causal value, the correlation coefficients, and time lags are shown in Table 3.2.

Comparing the results of Tables 3.1 and 3.2, we find that the percentile increases at the lowest two latitude circles (the 20.0° N and 22.5° N latitude circle) and decrease at other latitude circles (the 25.0° N–50.0° N latitude circle). The SAAs strongly affect the SATAs from the 20.0° N latitude circle to 32.5° N latitude circle, affect at the 35.0° N latitude circle, and do not affect from the 37.5° N to 50.0° N latitude circle.

The correlation coefficient increases at the lowest latitude circle. But the correlation coefficients decrease at other latitude circles, while all values are yet statistically significant over 98 % confidential level. The time lags −2 months at the lowest three latitude circles change to the reasonable value of 1 month. The time lag of the 37.5° N latitude circle changes from 5 months to 6 months. We can find from the above results that the ENSO warm event weakens or overwhelms the volcanic cooling at low latitudes and strengthens at midlatitudes and high latitudes. We also find that the maximum cooling occurs a few months after the month of the maximum solar insolation at each latitude. The linear interaction between the cooling effect of the volcanic sulfuric acid aerosols and the ENSO warm event makes it difficult to study the climate impact of volcanic eruptions in the troposphere. However, the present study proved that the net effect of volcanic aerosols on the global climate is rather simple, and it is clearly shown by removing the effect of the ENSO warm event from the SATAs.

3.4.3 Analysis of 3 Latitude Bands

The causal values from the SAAs to the SATAs are calculated for 3 latitude bands, and the percentile of the causal value is also obtained. They are shown in Table 3.3. The maximum negative correlation coefficients and their time lags are also calculated and are shown in the table. The SAAs strongly affect the SATAs of the 20–30° N and 30–40° N latitude bands, and affect the SATAs of the 40–50° N latitude band. The correlation coefficient is negative maximum at the 20–30° N latitude band, and the time lag n increases along with latitude.

Next, we remove the effect of the ENSO from the SATAs using the Nakano's formula [Nakano, 1999]. The percentile of the causal value from the SAAs to the SATAs removed the effect of the SOI is calculated. They are shown in Table 3.4 and the maximum negative correlation coefficients and their time lags are also shown in the table.

Comparing Tables 3.3 and 3.4, we find that the percentile of the causal value decreases at all latitude bands, the decreased amount is larger along with latitude, so that the percentile of the causal value decreases monotonically with latitude. The absolute value of correlation coefficients also decreases

Table 3.3: The percentile of the causal value from the SAAs to the SATAs and the correlation coefficients between the SAAs and the SATAs at 3 latitude bands.

latitude	correlation coefficient	time lag (month)	percentile of causal value
40–50° N	−0.5173	7	94.8
30–40° N	−0.6185	5	99.5
20–30° N	−0.6293	1	97.2

Table 3.4: The percentile of the causal value from the SAAs to the SATAs removed the effect of the SOI and the correlation coefficients between the SAAs and the SATAs removed the effect of the SOI at 3 latitude bands.

latitude	correlation coefficient	time lag (month)	percentile of causal value
40–50° N	−0.4330	7	87.2
30–40° N	−0.5363	5	90.3
20–30° N	−0.6047	2	96.2

along with latitude. A time lag n changes 2 from 1 at the 20–30° N latitude band. A time lag becomes large with latitude, and is a few months after the month of the maximum solar insolation. Namely, the maximum cooling due to volcanic aerosols follows the maximum solar insolation with a time lag of a few months. This result is consistent with the findings of Robock and Mao [1995], who described that the significant cooling region follows the location of the maximum solar insolation.

3.5 Discussion and Conclusions

The conclusions of this study are summarized as follows:

1. The magnitude of the Mt. Pinatubo eruption was so large that the evaluation of the volcanic impact on the climate was possible in spite of a single eruption. The direct measurement of the SAAs was performed for a long period including the time of the Mt. Pinatubo eruption. Then, we were able to make clear the cooling effect of the volcanic aerosols on the SATAs.

2. It is the common recognition from many studies so far that the SAAs become maximum around half a year after a volcanic eruption and the SATA response to volcanic impacts becomes prominent after a few years following volcanic eruptions. In the case of the Mt. Pinatubo eruption, the maximum SAAs occurred after eight months and the maximum cooling occurred after two years from the eruption. In this study, we removed the effect of the ENSO from the SATAs to make clear the pure effect of the SAAs on the SATAs. We found that the maximum cooling of the SATAs occurs at the time when the solar insolation is maximum in the next summer with a lag of a few months. It is found that the cooling amount of the SATAs is larger at lower latitudes, which is reasonable if the SAAs is almost uniform all over the globe, as it was in the Mt. Pinatubo case.

3. In reality, the maximum cooling of the SATAs occurred in the second (1993) summer, which was caused by the interaction between the SAAs and the El Niño which was occurring in 1993. The El Niño warmed the SATAs at low latitudes and enhanced the cooling at midlatitudes and high latitudes.

4. The causal analysis and the correlation analysis are performed for the SATAs removed the ENSO signal using the Nakano's formula [Nakano, 1999]. The results show that the SATAs of low latitudes are more significantly affected by the SAAs. The effect of the SAAs and the ENSO warm event in 1993 cancel each other in low latitudes and enhance cooperatively in midlatitudes and high latitudes to cool the earth surface. It is this linear interaction that brought a cool summer to the midlatitudes of the Northern Hemisphere in 1993.

5. It is shown in this study that the causal analysis [Nakano, 1995] and the Nakano's formula [Nakano, 1999] bring a new insight into the time sequential data analysis.

Chapter 4

The Effect of the ENSO on the Global Climate

by Osamu Morita

In this chapter, we will show the effect of the El Niño and Southern Oscillation (ENSO) phenomena on the global climate parameters, the air temperature at 1000 hPa level, the geopotential height of 1000 hPa surface, and the surface precipitation, using the causal analysis.

4.1 Introduction

It is a worldwide consensus that ENSO events affect the global climate. A several circum-Pacific countries make their own criteria to judge the occurrence of an El Niño to make their long-range weather forecast accurate. Because it is well known how and where the extreme weather occurs in the phase of El Niños. Extreme weather occurs not only in the tropics but also at regions far from the equator, which is called the teleconnection pattern. In this study, we will show the spatial structure of the direct and teleconnection effect of El Niños on the global climate parameters.

4.2 Data Used for This Study

Data used for this study are all monthly values, which are;

1. The Southern Oscillation (SO) Index (SOI), which is downloaded from the website of National Centers for Environmental Information, National Oceanic and Atmospheric Administration (NOAA)[1].

[1] https://www.ncei.noaa.gov/access/monitoring/enso/soi

DOI:10.1201/9781003603429-4

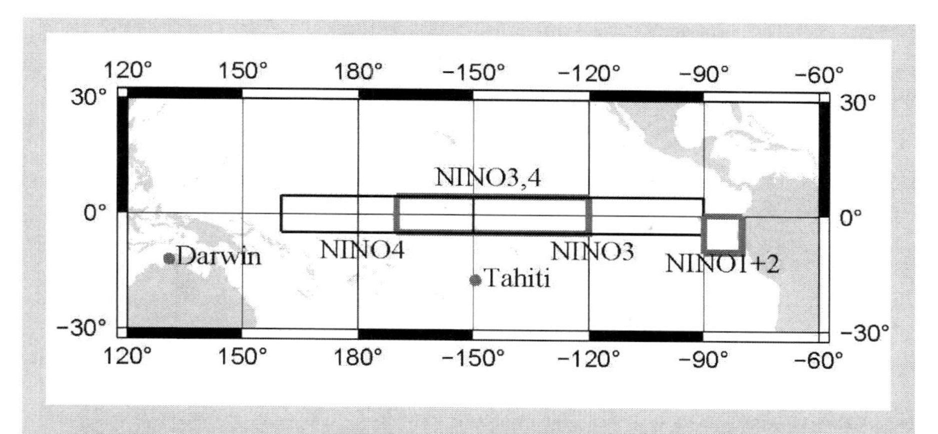

Figure 4.1: Locations of Darwin (130.8°E, 12.4°S), Tahiti (149.5°W, 17.6°S), and four Niño regions; Niño1+2 (0°–10°S, 90°W–80°W), Niño3 (5°N–5°S, 150°W–90°W), Niño3.4 (5°N–5°S, 170°W–120°W), and Niño4 (5°N–5°S, 160°E–150°W).

2. The North Atlantic Oscillation (NAO) Index (NAOI)[2].

3. The Arctic Oscillation (AO) Index (AOI), which is downloaded from the website of Climate Prediction Center, NOAA[3].

4. The sea surface temperature (SST) anomalies (SSTAs) of four Niño regions, which are provided by the Physical Sciences Laboratory (PSL), NOAA, Boulder, Colorado, USA[4] [Reynolds et al., 2002] (Figure 4.1).

5. The air temperature at the 1000 hPa level (SAT), and the geopotential height of the 1000 hPa surface (GPH) of the NCEP/NCAR Reanalysis 1 data[5] [Kalnay et al., 1996].

6. The surface precipitation (SPR) of CMAP (CPC[6] Merged Analysis of Precipitation) data[7] [Xie and Arkin, 1997; Schneider et al., 2013].

The SOI is defined based on the sea level pressure (SLP) difference between Tahiti and Darwin [Ropelewski and Jones, 1987] (Figure 4.1). In the normal

[2]The NAO index is downloaded from the website of Climate Research Unit, University of East Anglia (https://crudata.uae.ac.uk/data/nao/).

[3]https://www.cpc.ncep.noaa.gov/products/precip/CWlink/daily_ao_index/ao.shtml

[4]https://psl.noaa.gov/data/timeseries/monthly/NINO12 etc.

[5]These data are provided by the PSL, NOAA, Boulder, Colorado, USA, from their website (https://psl.noaa.gov).

[6]Climate Prediction Center, NOAA.

[7]These data are provided by the PSL, NOAA, Boulder, Colorado, USA, from their website (https://psl.noaa.gov/data/gridded/data.cmap.html).

condition, the SST is warmer and the SLP is lower at the western tropical Pacific and vice versa at the central and eastern tropical Pacific. In the El Niño condition, the warm ocean water moves to east, so that the SST becomes warmer and the SLP becomes lower than the normal condition at the central tropical Pacific. Thus, the SOI is positive in the normal or La Niña[8] condition and negative in the El Niño condition.
The NiñoX (X is 1+2, 3, 3.4, and 4) SST anomalies (SSTAs) are calculated as follows:

1. The observed SST of the NiñoX region is averaged all over the region to make the actual SST.

2. The long-term (30 year) average of the actual SST is calculated for each month to make the SST climatologies.

3. The NiñoX SSTAs are obtained by subtracting the SST climatologies from the actual SST.

Hereafter, we will call the NiñoX SSTAs as the NiñoX Index and abbreviate NiñoXI. The SAT, and the GPH are given as the global gridded data of 144×73 points, whose data period is from January 1981 to December 2010. We calculate the long-term (30 year) mean for each month (climatologies), which are subtracted from the actual data to make anomalies. The anomalies of the SAT and the GPH are designated SATAs and GPHAs. We usually used the 144-month (from January 1999 to December 2010) data for this study, and the 360 month (from January 1981 to December 2010) data for the reference. The SPR are given as the global gridded data of 144×72 points, whose data period is from January 1981 to December 2010. We made the time series of the SPR anomalies (SPRAs) as the same procedure mentioned above. We used the 144-month (from January 1999 to December 2010) data for this study.

4.3 Causal Relations between Climate Indices

In this section, we will examine the causal relationship between the SO, the NAO, the AO, and the ENSO Indices.

4.3.1 The Causal Relation between the SO and the NAO

We will investigate the causal relationship between the NAO and the SO, between the NAO and the AO, and between the SO and the AO. In the calculation of the percentile of the causal value, we examined the dependence of the percentile on the data number.
In Table 4.1, we show the data number dependence of the percentile of the causal value between the SOI and the NAOI. The value of the percentile

[8]The SST of the western tropical Pacific is warmer than the normal condition.

Table 4.1: The percentile of the causal value between the NAOI and the SOI, showing the dependence of the percentile of the causal value on the data number.

month	from SOI	from NAOI
120	58.6	89.6
144	41.2	90.3
180	7.0	85.7
210	29.4	92.5
240	28.9	96.6
300	25.3	95.7
360	19.7	96.2

becomes stable or saturated when the data number exceeds 210 months. The minimum significant data number seems to depend on the signal-to-noise ratio, the characteristic period of the objective phenomena, and so on.
We find that the NAOI affects the SOI but the SOI does not affect the NAOI at all. There is no causal relationship between the SOI and the AO Index (AOI), and between the NAOI and the AOI (results are not shown).

4.3.2 Causal Relation between the SOI and NiñoXI

We will investigate the percentile of the causal value between the SOI and four NiñoXI. The result is summarized in Table 4.2. The SOI intensely affects the Niño1+2I, Niño3I, and Niño4I, but does not affect the Niño3.4I. In the period from January 1999 to December 2010, there occurred four moderate El Niños. In the case of the moderate El Niño, the eastward progress of the warm oceanic water mass stops at the central equatorial Pacific (near Tahiti),

Table 4.2: The percentile of the causal value from the SOI to other ENSO Indices.

	from SOI
Niño1+2I	100.0
Niño3I	97.5
Niño3.4I	69.2
Niño4I	97.9

Table 4.3: The percentile of the causal value from the Niño1+2I to other ENSO Indices.

	from Niño1+2I
SOI	88.9
Niño3I	28.6
Niño3.4I	51.6
Niño4I	79.6

Table 4.4: The percentile of the causal value from the Niño3I to other ENSO Indices.

	from Niño3I
SOI	99.8
Niño1+2I	99.8
Niño3.4I	46.6
Niño4I	94.3

Table 4.5: The percentile of the causal value from the Niño3.4I to other ENSO Indices.

	from Niño3.4I
SOI	100.0
Niño1+2I	93.9
Niño3I	75.8
Niño4I	100.0

Table 4.6: The percentile of the causal value from the Niño4I to other ENSO Indices.

	from Niño4I
SOI	100.0
Niño1+2I	81.5
Niño3I	38.8
Niño3.4I	69.2

where the updraft of the Walker circulation occurs. The Niño3.4 region locates near the center of the Walker circulation, so that the wind stress converges sea surface water from the Niño3 region and the Niño4 region, and makes the SSTA fluctuation of the Niño3.4 region small.
The Niño4I, Niño3.4I, and Niño3I strongly affect the SOI, but the Niño1+2I does not affect the SOI. This result seems reasonable, because the Niño1+2 region is far from Tahiti and Darwin, and the area of this region is very small.

4.4 Effect of the ENSO Indices on the Global Climate

In this section, we will show the cause-and-effect relationship between five ENSO Indices and three global climate parameters using the causal analysis[9]

[9] In the causality theory [Nakano, 1995], it is judged that when the percentile of the causal value from a time series $\mathcal{Y}$ to a time series $\mathcal{X}$ is higher than 95, $\mathcal{Y}$ strongly affects $\mathcal{X}$, and when the percentile is between 90 and 95, $\mathcal{Y}$ affects $\mathcal{X}$.

During the period from January 1999 to December 2010, four moderate El Niño events (2003–2003, 2004–2005, 2006–2007, and 2009–2010) occurred.

4.4.1 Effect of the SO on the Global Climate Parameters

In this subsection, the effect of the SO on the SATAs, GPHAs, and SPRAs are investigated for the global grid points. The results are shown in the following.

(A) Effect of the SO on the Global SATAs

The percentile of the causal value from the SOI to the SATAs is computed for the global 144×73 grid points. The results are shown in Figure 4.2(a), which is the conformal Mercator projection from 80°N to 80°S centering at 180°E and 0° and shows the global structure of the percentile of the causal value from the SOI to the global SATAs.

In the tropics, the SO affects the Atlantic Ocean, Central America, tropical South America, the Pacific, the Bay of Bengal, the Indian Ocean, the Indian subcontinent, the southern Arabian Peninsula, and the central Africa. It is remarkable that the nonaffected region extends at the central region of the tropical Pacific, which may be the node of the SO.

In the northern midlatitudes, the SO affects the central and eastern North Pacific and the west coast of United States. In the southern midlatitudes, the SO affects the central South Pacific, the southern Indian Ocean, and the west coast of Australia. These affected regions seem to make teleconnection patterns, namely the wave trains of the stationary Rossby waves [Held et al., 2002]. The teleconnections were first found by Walker [1923, 1924] for the time series of the air pressure, temperature and precipitation. Later, they were studied for surface air temperature anomalies (SATAs) and SPRAs [Seager et al., 2003; Seager et al., 2005; Ramadan et al., 2011].

In the northern high latitudes and within the Arctic Circle, the affected regions are the west coast and offshore of Greenland, the offshore of eastern Greenland, and the coast and offshore of northwestern Russia. We can't find the teleconnection patterns from the equator to these regions. In the southern high latitudes and around Antarctica, the intensely affected regions are southern South America, the Bellingshausen Sea, and the King Haakon VII Sea.

(B) Effect of the SO on the Global GPHAs

The percentile of the causal value from the SOI to the GPHAs is computed for the global 144×73 grid points. The results are shown in Figure 4.2(b).

In the tropics, the SO affects the equatorial Atlantic Ocean, the Gulf of Mexico, the eastern and western tropical Pacific, Southeast Asia, the Bay of Bengal, the equatorial Indian Ocean, Madagascar, and tropical Africa. The nonaffected region of the central Pacific is the intermediate area between

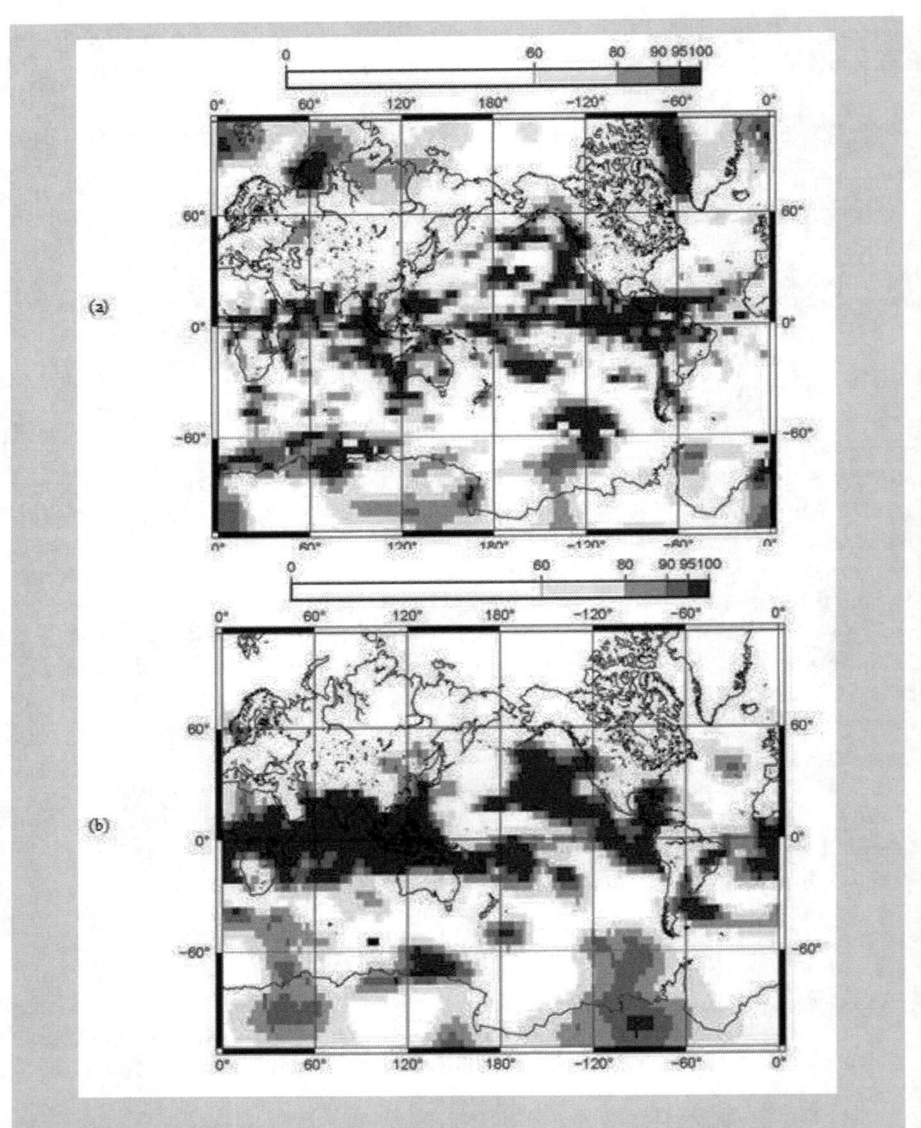

Figure 4.2: The spatial structure of the percentile of the causal value from the SOI to the SATAs (upper panel), and from the SOI to the GPHAs (lower panel) on the conformal Mercator projection (80°N–80°S) centering at (0°, 180°E).

Tahiti and Darwin. There is a possibility that the area is the node of the SO.

In the northern midlatitudes, the affected regions are the North Pacific, southwestern Japan, the East China Sea, and northern India. It is not

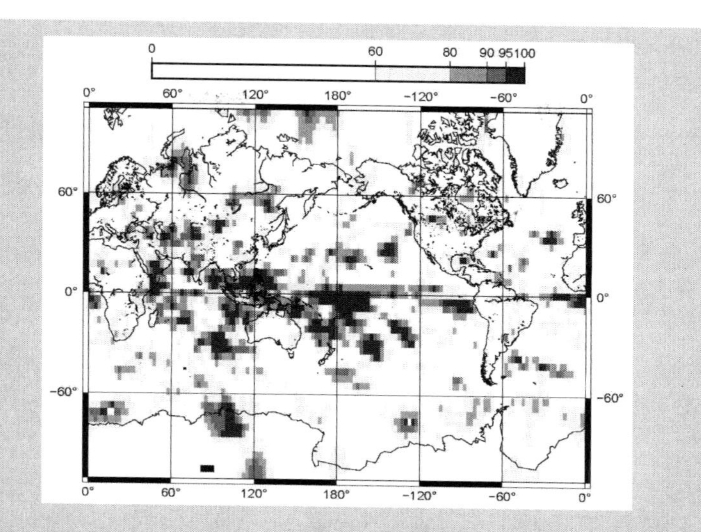

Figure 4.3: The spatial structure of the percentile of the causal value from the SOI to the SPRAs on the conformal Mercator projection (80°N–80°S) centering at (0°, 180°E).

obvious whether these regions are directly affected by the SO or the results of teleconnections. In the southern midlatitudes, there are two teleconnection patterns, which seem to reach Antarctica.

In the southern high latitudes and around Antarctica, the affected regions are the eastern coast and offshore of Argentine, the belt area from the Weddel Sea to the South Pole, and the coast and offshore of George V Land.

(C) Effect of the SO on the Global SPRAs

The percentile of the causal values from the SOI to the SPRAs is computed for the global 144×72 grid points. The results are shown in Figure 4.3, which are the same as Figure 4.2 except for the spatial structure of the percentile of the causal value from the SOI to the SPRAs.

In the tropics, the SO affects the SPRAs at the equatorial Atlantic Ocean, the Niño1+2 region, the tropical Pacific, the western equatorial Pacific, Southeast Asia, the Bay of Bengal, southwestern India, the southern Arabian Peninsula, Madagascar, and eastern Africa.

In the northern midlatitudes, the affected regions are sparse and small at the central Pacific, the central Atlantic Ocean, and the Eurasian Continent. In the southern midlatitudes, the affected regions are small and sparse at the central Pacific, the Bay of Bengal, and the southern Indian Ocean.

In the northern high latitudes and within the Arctic Circle, the affected regions are northwestern Russia, and around the North Pole. In the southern high latitudes and around the Antarctica, the affected regions are the Bellingshausen See, and Princess Elizabeth Land which are small and sparse.

4.4.2 Effect of the Niño1+2I on the Global Climate Parameters

In this subsection the effect of the Niño1+2I on the SATAs, GPHAs, and SPRAs are investigated for the global grid points. The results are shown in the following.

(A) Effect of the Niño1+2I on the Global SATAs

The percentile of the causal values from the Niño1+2I to the SATAs is computed for the global 144×73 grid points, whose result is shown in Figure 4.4(a).
In the tropics, the Niño1+2I affects tropical South America, the central and eastern tropical Pacific, Indonesia, Southeast Asia, the Indian Subcontinent, the equatorial Indian Ocean, Madagascar, and tropical Africa.
In the northern midlatitudes, the Niño1+2I affects the offshore of eastern United States, the North Pacific where areas are small and sparse, and the eastern Mediterranean Sea. In the southern midlatitudes, the Niño1+2I affects the South Atlantic Ocean, the South Pacific where regions are small and sparse, southeastern Australia, and the southern Indian Ocean.
In the northern high latitudes and within the Arctic Circle, the affected regions are northwestern Russia, and northern Fennoscandia. In the southern high latitudes and around Antarctica, the affected regions are Wilkes Land, and Enderby Land.

(B) Effect of the Niño1+2I on the Global GPHAs

The percentile of the causal value from the Niño1+2I to the GPHAs is computed for the global 144×73 grid points. The results are shown in Figure 4.4(b).
In the tropics, the Niño1+2I affects the western coast of Africa, the tropical Atlantic Ocean, tropical South America, the eastern tropical Pacific, the western tropical Pacific, Southeast Asia, the Bay of Bengal, the southern Indian Subcontinent, the tropical Indian Ocean, and equatorial Africa.
In the northern midlatitudes, the affected regions are the offshore of the eastern Labrador Peninsula, northwestern United States, the central and western Pacific, the central Eurasian Continent, and western Europe.
In the northern high latitudes and within the Arctic Circle, the offshore of northwestern Russia is affected. In the southern high latitudes and around Antarctica, the affected regions are the Strait of Magellan, the Ross Ice Shelf, and George V Land.

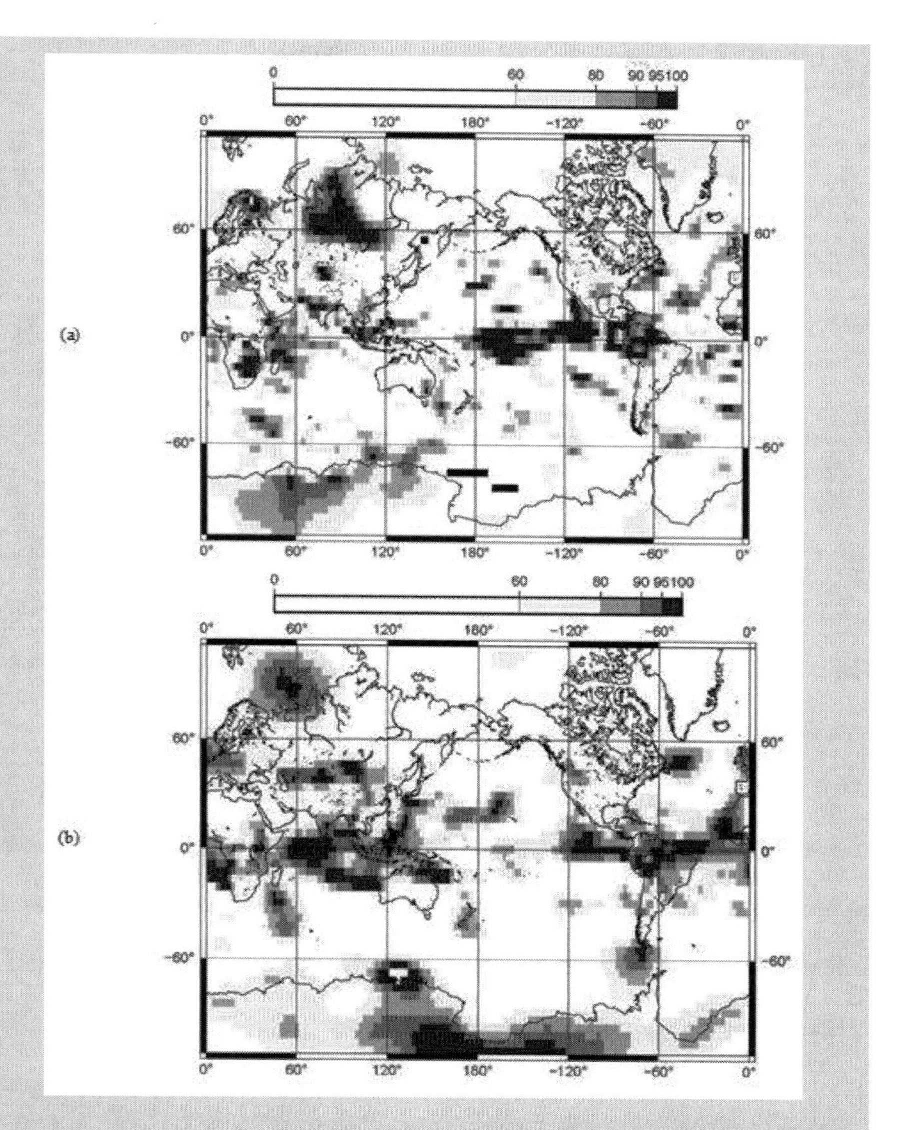

Figure 4.4: The spatial structure of the percentile of the causal value from the Niño1+2I to the SATAs (upper panel) and from the Niño1+2I to the GPHAs (lower panel) on the conformal Mercator projection (80°N–80°S) centering at (0°, 180°E).

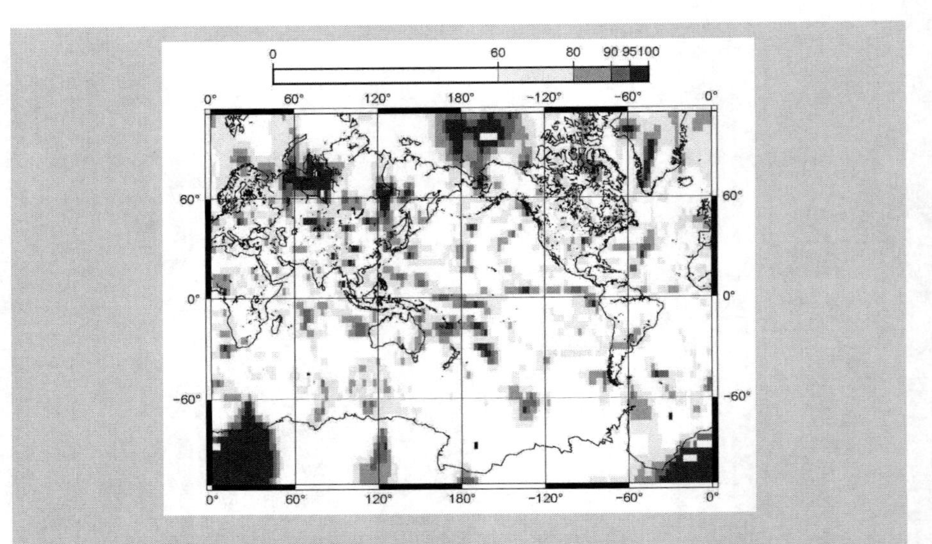

Figure 4.5: The spatial structure of the percentile of the causal value from the Niño1+2I to the SPRAs on the conformal Mercator projection (80°N–80°S) centering at (0°, 180°E).

(C) Effect of the Niño1+2I on the Global SPRAs

The percentile of the causal values from the Niño1+2I to the SPRAs is computed for the global 144×72 grid points. The results are shown in Figure 4.5. In the tropics, the Niño1+2I affects the SPRAs at the tropical Pacific, the tropical Indian Ocean, northern Madagascar, and the west coast of equatorial Africa, where the affected regions are sparse and non-systematic.

In the northern midlatitudes, the affected regions are the North Atlantic Ocean where the affected regions are sparse, the North Pacific, and the Eurasian Continent. In the southern midlatitudes, the affected regions are South America, the South Pacific, the Indian Ocean, and the southwest sea of Cape Town, which are sparse and small.

In the northern high latitudes and within the Arctic Circle, the affected regions are northern and northwestern Russia, and the northern sea of Alaska. In the southern high latitudes and around Antarctica, affected regions are the Southern Ocean, and MacRobertson Land which are very sparse.

4.4.3 Effect of the Niño3I on the Global Climate Parameters

In this subsection the effect of the Niño3I on the SATAs, GPHAs, and SPRAs are investigated for the global grid points. The results are shown in the following.

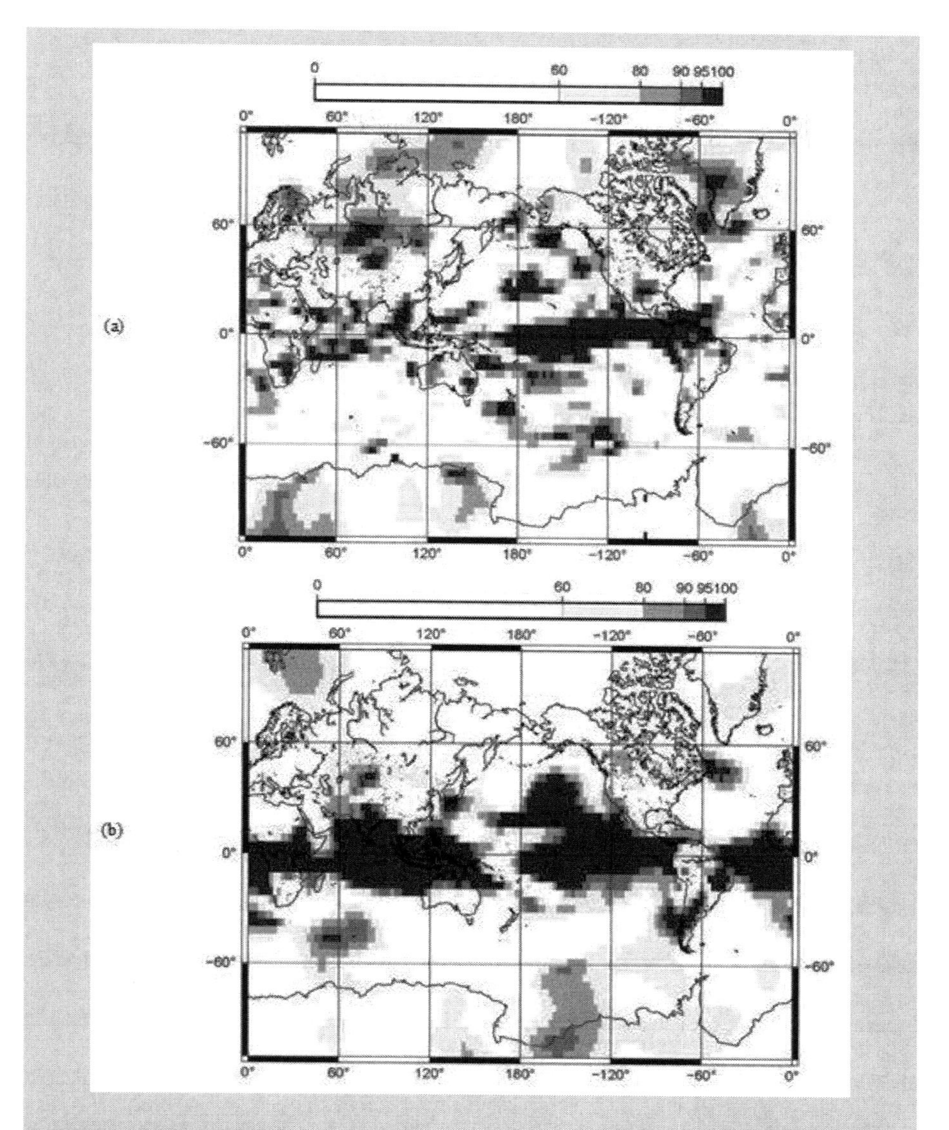

Figure 4.6: The spatial structure of the percentile of the causal value from the Niño3I to the SATAs (upper panel) and from the Niño3I to the GPHAs (lower panel) on the conformal Mercator projection (80°N–80°S) centering at (0°, 180°E).

(A) Effect of the Niño3I on the Global SATAs

The percentile of the causal value from the Niño3I to the SATAs is computed for the global 144×73 grid points. The results are shown in Figure 4.6(a).

In the tropics, the Niño3I affects the western tropical Atlantic Ocean, northern South America, United States facing the Gulf of Mexico, the Niño1+2 region, the central and eastern tropical Pacific, which form a wide and thick belt. Further, the affected regions are at Southeast Asia, the tropical Indian Ocean, Madagascar, the southern Arabian Peninsula, and equatorial Africa.
In the northern midlatitudes, the Niño3I affects the northeastern and central North Pacific, which seem to form a teleconnection pattern. Further, there are affected regions at the central Eurasian Continent, and southern Europe. In the southern midlatitudes, the Niño3I affects southeastern and western Australia, the central South Pacific, and South Africa.
In the northern high latitudes and within the Arctic Circle, the affected regions are the northern North Atlantic, the western coast and offshore of Greenland, the northwestern North Pacific, northern and northeastern Russia, the central Eurasian Continent, and northeastern Fennoscandia. In the southern high latitudes and around Antarctica, the affected regions are the southern South Pacific, Oates Land, and Kemp Land.

(B) Effect of the Niño3I on the Global GPHAs

The percentile of the causal value from the Niño3I to the GPHAs is computed for 144×73 grid points over the globe. The results are shown in Figure 4.6(b). In the tropics, the Niño3I affects the GPHAs around the equator except for northern South America and the Niño4 region.
In the northern midlatitudes, the affected regions are the coast and offshore of the Labrador Peninsula, the central North Pacific, southern Japan, and eastern Uzbekistan. In the southern midlatitudes, the affected regions are southern South America, the southern Indian Ocean, and the southern sea of South Africa.
In the southern high latitudes and around Antarctica, the affected regions are around the Strait of Magellan, and the Ross Iceshelf.

(C) Effect of the Niño3I on the Global SPRAs

The percentile of the causal values from the Niño3I to the SPRAs is computed for the global 144×72 grid points. The results are shown in Figure 4.7.
In the tropics, the Niño3 Index affects the SPRAs at the tropical Pacific, Southeast Asia, the northwestern coast and offshore of Australia, the Bay of Bengal, the northern Indian Ocean, Madagascar, and equatorial Africa.
In the northern midlatitudes, the affected regions are the North Atlantic where they are small and sparse, the North Pacific, and the Eurasian Continent. In the southern midlatitudes, the affected regions are southern South America where they are sparse, the South Pacific, the southern Indian Ocean, and South Africa.
In the northern high latitudes and within the Arctic Circle, the affected regions are southwestern Greenland, the offshore of the Arctic coast of Alaska, and

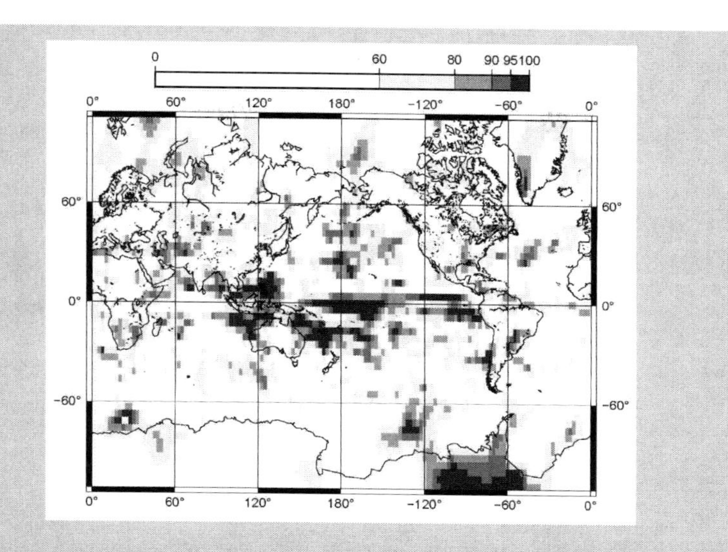

Figure 4.7: The spatial structure of the percentile of the causal value from the Niño3I to the SPRAs on the conformal Mercator projection (80°N–80°S) centering at 180°E, The spatial structure of the percentile of the causal value from the Niño3I to the SPRAs on the conformal Mercator projection (80°N–80°S) centering at (0°, 180°E).

Novaya Zemlya Islands of Russia. In the southern high latitudes and around Antarctica, affected regions are the Bellingshausen Sea, and the base of the Antarctic Peninsula.

4.4.4 Effect of the Niño3.4I on the Global Climate Parameters

In this subsection the effect of the Niño3.4I on the SATAs, GPHAs, and SPRAs are computed for the global grid points. The results are shown in the following.

(A) Effect of the Niño3.4I on the Global SATAs

The percentile of the causal value from the Niño3.4I to the SATAs is computed for the global 144×73 grid points. The results are shown in Figure 4.8(a).
In the tropics, the Niño3.4I affects western Africa, the tropical North Atlantic Ocean, northern South America, the coast of United States facing the Gulf of Mexico, Central America, the tropical Pacific which form a wide and continuous belt. Further, the affected regions are Southeast Asia, the tropical Indian Ocean, Madagascar, the southern Arabian Peninsula, and equatorial Africa.

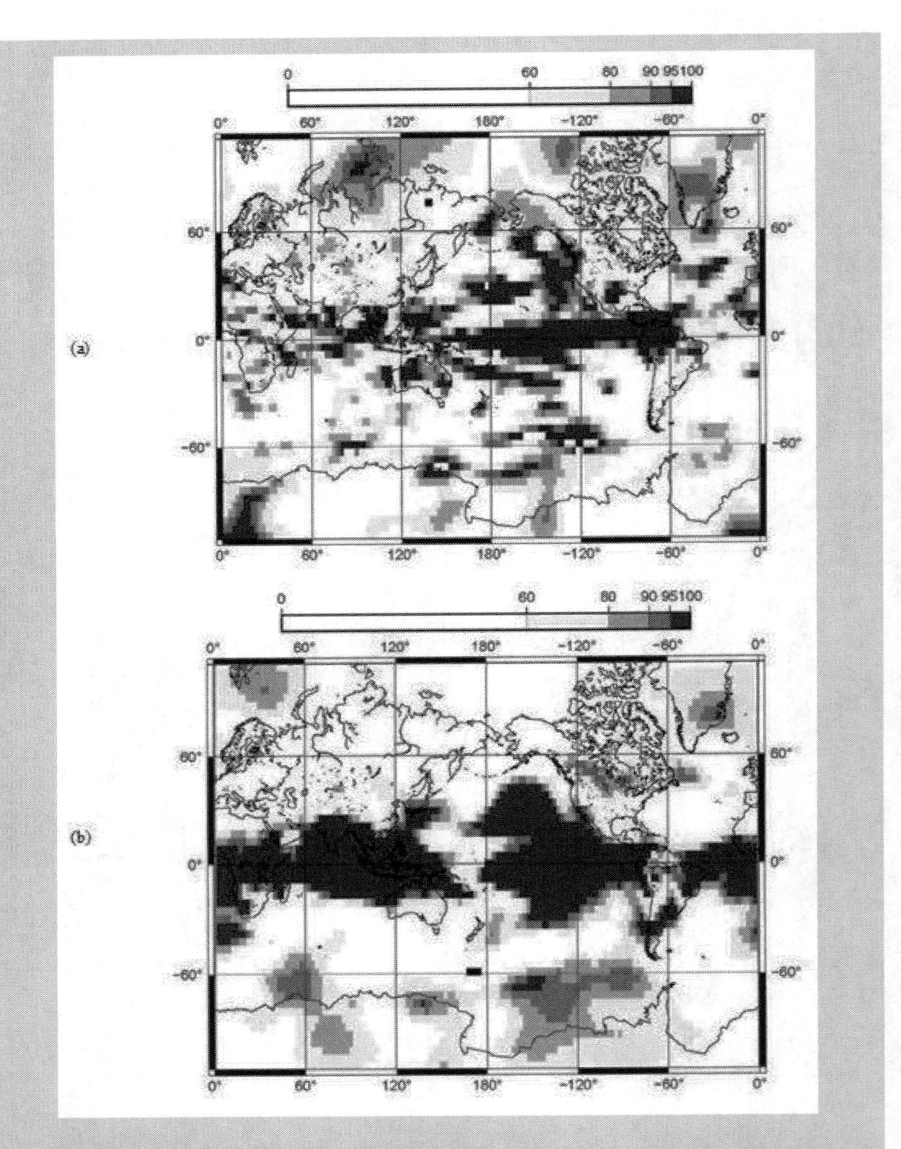

Figure 4.8: The spatial structure of the percentile of the causal value from the Niño3.4I to the SATAs (upper panel) and from the Niño3.4I to the GPHs (lower panel) on the conformal Mercator projection (80°N–80°S) centering at (0°, 180°E).

In the northern midlatitudes, the Niño3.4I affects the northeastern and central North Pacific, which seem to form a teleconnection pattern. Further, there are affected regions at the central North Atlantic Ocean, and the Mediterranean Sea. In the southern midlatitudes, there is an affected belt in the South Pacific which looks like a teleconnection pattern. In addition, affected regions are New Zealand, southeast central and western Australia, and the coast and offshore of South Africa.
In the northern high latitudes and within the Arctic Circle, the affected regions are southern Greenland, the offshore of southeastern Greenland, and the offshore and the northern coast of Russia. In the southern high latitudes and around Antarctica, the affected regions are two zonal belts, one is the Ross Sea and the other is Dronning Maud Land.
The pattern of Figure 4.8(a) resembles that of Figure 4.7(a), which is the effect of the Niño3I on the global SATAs.

(B) Effect of the Niño3.4I on the Global GPHAs

The percentile of the causal value from the Niño3.4I to the GPHAs is computed for 144×73 global grid points. The results are shown in Figure 4.5(b). In the tropics, the Niño3.4I affects the GPHAs all over the tropics except northern South America, and the Niño4 region and its northern sea. The reason why the GPHAs of Niño4 region is not affected by the Niño3.4I may be that the Niño4 region and its northern sea is the node of the SO.
In the northern midlatitudes, the affected regions are the east coast and offshore of the Labrador Peninsula, the northwest-southeast belt structure in western Canada and northern United States, the central North Pacific, the southern sea of Japan. In the southern midlatitudes, the affected regions are the southern sea of South Africa, and southern South America.
In the northern high latitudes and within the Arctic Circle, the affected regions are the northern sea of the Scandinavian Peninsula, and southeastern Greenland. In the southern high latitudes and around Antarctica, the affected regions are the Bellingshausen Sea, and the coast and offshore of Queen Mary Land.
It is worth noticing that the spatial structure of Figure 4.8(b) resembles that of Figure 4.7(b).

(C) Effect of the Niño3.4I on the Global SPRAs

The percentile of the causal values from the Niño3.4I to the SPRAs is computed for the global 144×72 grid points. The results are shown in Figure 4.9. In the tropics, the Niño3.4I affects the SPRAs at the tropical Pacific, Southeast Asia, the northeastern coast and offshore, and the northwest coast and offshore of Australia, Madagascar, and the tropical Indian Ocean.
In the northern midlatitudes, the affected regions are the North Atlantic Ocean where they are small and sparse, the central North Pacific, and a zonal

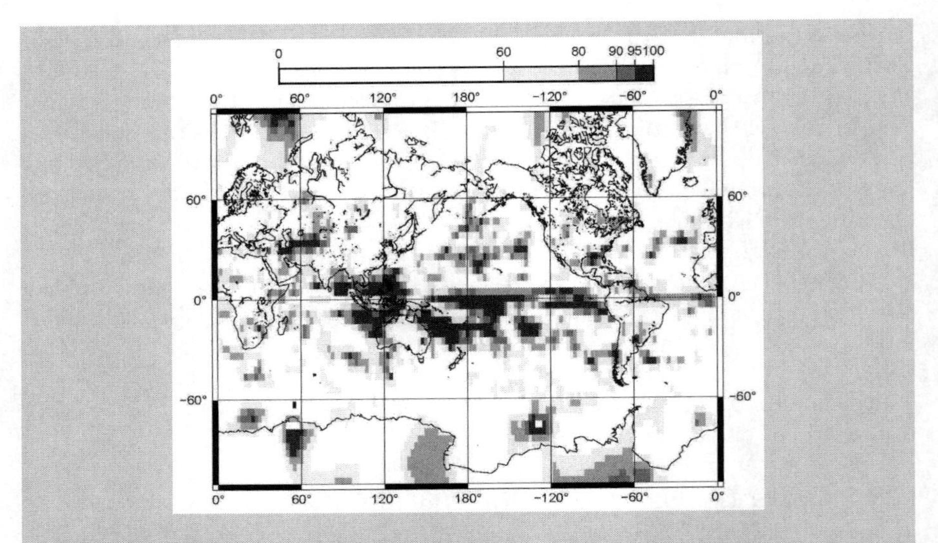

Figure 4.9: The spatial structure of the percentile of the causal value from the Niño3.4I to the SPRAs on the conformal Mercator projection (80°N–80°S) centering at (0°, 180°E).

belt structure east of the Caspian Sea. In the southern midlatitudes, affected regions are the east coast and offshore of Argentine, a zonal belt in the central South Pacific, the southern Indian Ocean, Madagascar, and eastern South Africa.

In the northern high latitudes and within the Arctic Circle, the affected regions are the east coast of Greenland, and the northwestern sea of Novaya Zemlya Islands of Russia. In the southern high latitudes and around Antarctica, affected regions are the base of the Antarctic Peninsula, the Bellingshausen Sea, and Kemp Land of Antarctica.

4.4.5 Effect of the Niño4I on the Global Climate Parameters

In this subsection the effect of the Niño4I on the SATAs, GPHAs, and SPRAs are examined for the global grid points. The results are shown in the following.

(A) Effect of the Niño4I on the Global SATAs

The percentile of the causal value from the Niño4I to the SATAs is computed for the global 144×73 grid points. The results are shown in Figure 4.6(b).

In the tropics, the Niño4I affects Central America, tropical South America, the tropical Pacific, Australia, Southeast Asia, the tropical Indian Ocean, southern India, the southern Arabian Peninsula, Madagascar, and tropical Africa. In the northern midlatitudes, the Niño4I affects the northern North Atlantic Ocean, the northeastern and central North Pacific, and southern Europe. In the southern midlatitudes, affected regions are the east coast of South America, the southern South Pacific, Australia, Madagascar, and the southwestern sea of South Africa.

In the northern high latitudes and within the Arctic Circle, the affected regions are central Greenland, and northern Russia. In the southern high latitudes and around Antarctica, the affected regions are the Antarctic Peninsula, the Ross Sea, the coast and offshore of George V Land, and the offshore of Prydz Bay of Antarctica.

The spatial structure of Figure 4.10(a) resembles that of Figure 4.9(a) which is the effect of the Niño3.4I on the SATAs.

(B) Effect of the Niño4I on the Global GPHAs

The percentile of the causal value from the Niño4I to the GPHAs is computed for 144×73 global grid points. The results are shown in Figure 4.10(b).

In the tropics, the Niño4 Index affects the GPHAs all over the tropics except the Niño4 region and its northern sea. The reason why this area is not affected by the Niño4I may be that the area is the node of the SO.

In the northern midlatitudes, the affected regions are the eastern and the central North Pacific, southern Japan, the southern sea of Japan, and the East China Sea. In the southern midlatitudes, the affected regions are southern Argentine, the southern South Pacific, and the west coast and the offshore of South Africa.

In the southern high latitudes and around Antarctica, the affected regions are the Bellingshausen Sea and the coast facing it, the coast of George V Land, and the coast and the offshore of Kemp Land.

The affected regions of the Southern Hemisphere look like to form the teleconnection patterns. It is worth noticing that the pattern of Figure 4.10(b) resembles those of Figures 4.8(b) and 4.9(b). And it is interesting that we cannot find any teleconnection pattern in the Northern Hemisphere.

(C) Effect of the Niño4I on the Global SPRAs

The percentile of the causal value from the Niño4I to the SPRAs is computed for the global 144×72 grid points. The results are shown in Figure 4.11.

In the tropics, the Niño4I affects the SPRAs at the equatorial Atlantic Ocean, the tropical Pacific, Southeast Asia, northern Australia, the Bay of Bengal, the tropical Indian Ocean, northern Madagascar, and equatorial Africa.

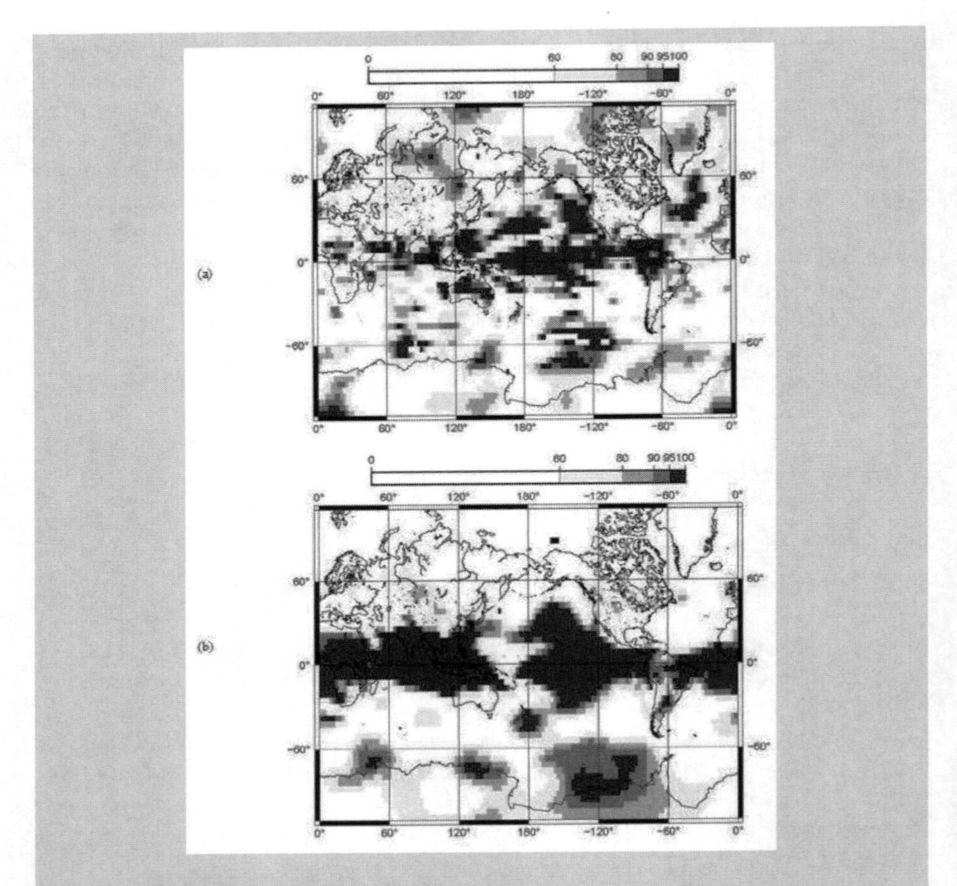

Figure 4.10: The spatial structure of the percentile of the causal value from the Niño4I to the SATAs (upper panel) and from the Niño4I to the GPHAs (lower panel) on the conformal Mercator projection (80°N–80°S) centering at (0°, 180°E).

In the northern midlatitudes, the affected regions are the North Pacific where they are small and sparse, southwestern Japan, and the belt area from Iraq to southern Kazakhstan. In the southern midlatitudes, affected regions are Argentine, the central South Pacific, eastern Australia, and the southern Indian Ocean, which are in the form of northwest-southeast belt structures.
In the southern high latitudes and around Antarctica, affected regions are the Bellingshausen Sea, the coast of Oates Land, and Kemp Land.

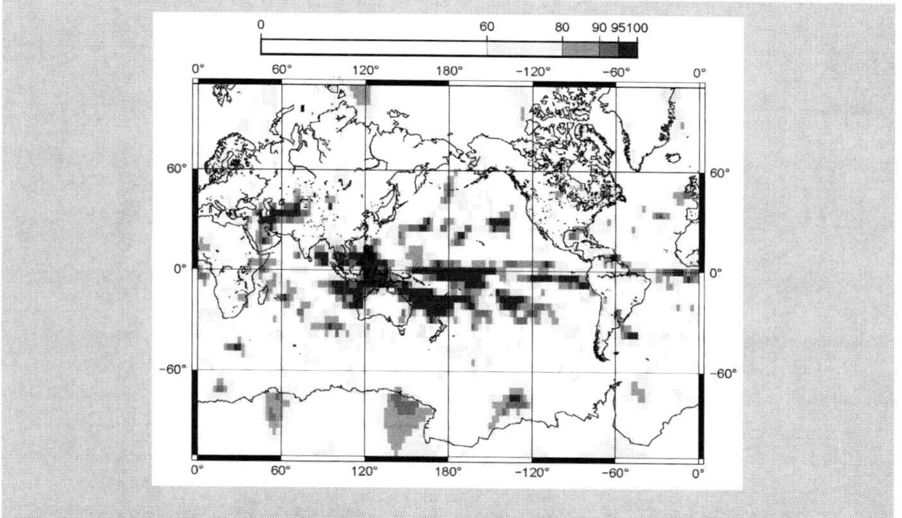

Figure 4.11: The spatial structure of the percentile of the causal value from the Niño4I to the SPRAs on the conformal Mercator projection (80°N–80°S) centering at (0°, 180°E).

4.5 Discussion and Summary

4.5.1 As for the Data Period

The causality analysis is designed to treat a time sequential data of data number over 100. It has already been explained in subsection 4.3.1 that the data number is crucial in the causality analysis. In this chapter, most analyses are performed using the data number of 144 months, and we will show the reason why we use the data number 144.
Figure 4.12 shows the spatial structure of the percentile of the causal value from the SOI to the GPHAs for the data period of January 1981–December 2010 (360 months). Comparing this figure with Figure 4.2(b), we find that the affected area of the former is broader and the fine structure is more obscure than those of the latter. So, we adopted the data number of 144 months (January 1999–December 2010) in this study.

4.5.2 Effect of the SO

The SOI affects the SATAs, GPHAs, and SPRAs all over the globe. The structure of affected areas from the tropics to the extratropics shows teleconnection patterns which were found by Walker [1923, 1924] and now are understood as the wave trains of stationary Rossby waves [Held et al., 2002]. Comparing

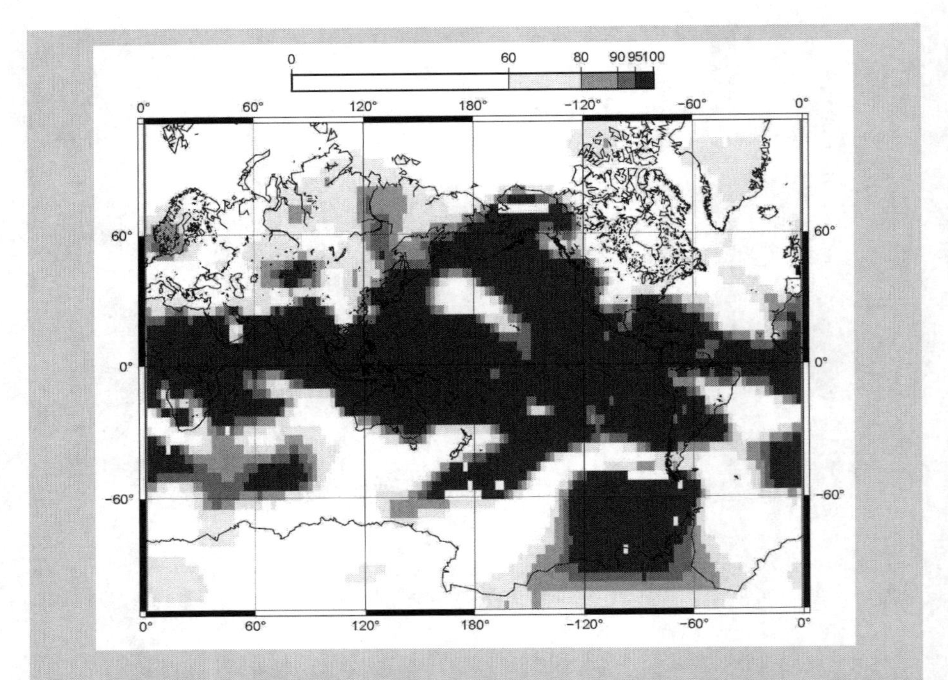

Figure 4.12: Same as Figure 4.2(b) except the data period is from January 1981 to December 2010 (360 month).

the effect of the SO on three climate parameters, the effect on the GPHAs is more intense than the effect on SATAs and SPRAs. The effect of the SO on the SATAs and SPRAs extend to midlatitudes and high latitudes of the both hemispheres, while the effect of SO on the GPHAs extends to midlatitudes of the both hemispheres but does not to high latitudes of the Northern Hemisphere. We can't find yet the mechanism of the above phenomenon, and leave it for future study.

4.5.3 Effect of Four NiñoXI

The structures of affected regions of three climate parameters by the Niño3I, Niño3.4I, and Niño4I resemble each other, while the structure by the Niño1+2I is fairly different. The reason is that the area of Niño1+2 is very small and its position is the east end of the tropical Pacific. Further, four El Niños which occurred in the analyzing period were all moderate one, in which the eastward progress of the warm SSTA stopped at the center of the tropical Pacific.
The SOI, the Niño3I, Niño3.4I, and Niño4I affect the SATAs, GPHAs, and SPRAs all over the globe. The structure of affected areas from the tropics to the extratropics, which seem to be the direct influence and the teleconnection

patterns. The affected regions of the Niño1+2I on the three climate parameters also extend all over the globe. However, the influence of the Niño1+2I is far weaker than that of other ENSO Indices.
Among the effect of ENSO Indices on three climate parameters, the effect on the GPHAs is more intense than on the SATAs and SPRAs. The effect on the SATAs and SPRAs extend to midlatitudes and high latitudes of the both hemispheres, while the effect on the GPHAs extends to midlatitudes of the both hemispheres but does not to high latitudes of the Northern Hemisphere and within the Arctic circle.

4.5.4 Possible Mechanism of the Effect of the NAO on the SO

We show in subsection 4.3.1 that the NAO affects the SO, which is the important finding in this study. We will consider the possible mechanism how the NAO affects the SO. Figure 6.2 is the spatial structure of the percentile of the causal value from the NAOI to the GPHAs of 144×73 global grid points. The data period is from January 1981 to December 2010 (240 months).
In the tropics, the NAO affects the GPHAs of the Niño3 region and its adjacent northern and southern seas, the central South Pacific, the western sea of Australia, the western tropical Pacific, the southern Arabian Peninsula, Madagascar and its eastern sea, and tropical Africa.
In the northern midlatitudes, the affected regions are the central Atlantic Ocean, northeastern United States, southern Japan, and northern Europe. In the southern midlatitudes, the affected regions are the eastern sea of Argentine, the western sea of Chile, the central South Pacific, northern Australia, the eastern sea of Madagascar, and Angola.
In the northern high latitudes and within the Arctic Circle, the affected regions are southeastern Fennoscandia, northeastern Europe, western Russia. In the southern high latitudes and around Antarctica, the affected regions are the southern sea of South America, the eastern and western sea of the Antarctic Peninsula, and the Bellingshausen Sea.
It is very interesting that the NAO affects the GPHAs of not only the North Atlantic Ocean and its periphery regions but also the Pacific, Eurasian Continent, and the Southern Hemisphere. The effect on the Southern Hemisphere may be caused indirectly through the SO.
Hurrell et al. [2003] state that in the winter season (December–March) of the large positive NAO Index (> 1.0), the $\mathbb{SAT}$As extend through the Eurasian continent from the North Atlantic to the North Pacific and United States, and the negative $\mathbb{SAT}$As prevail in eastern Canada, northern Africa, and the Arabian Peninsula. The spatial distribution of the $\mathbb{SAT}$As causes the SLP anomalies (SLPAs), and induces the atmospheric circulation.
In our study, it is shown another possible mechanism that the SLPAs of Azores High may cause the SLPAs of the tropical Pacific, which can be a trigger of starting the EL Niño.

Chapter 5

The Effect of ENSO on the Japanese Climate

by Osamu Morita

In this chapter, we will consider the effect of El Niño and Southern Oscillation (ENSO) on the Japanese climate parameters, the surface air temperature, the sea level pressure, and the surface precipitation, using the causal analysis.

5.1 Introduction

In Japan, it is well known that we have a cool summer and a warm winter, and a long-term rainy season (Baiu) in the period of El Niño (the warm phases of ENSO) and vice versa in the period of La Niña. Maeda [2013] states that the influence of ENSO on Japanese climate[1] as follows:

1. The correlation coefficient between the Niño3 sea surface temperature (SST) deviation and the regional averaged surface air temperature (RSAT) is under 0.6, i.e. the relationship between the Niño3 SST deviation and the RSAT of Japan is not so strong. The effect of the Niño3 SST deviation on the RSAT is weaker in northern Japan than in other districts. During the period of El Niño, the RSAT anomalies (RSATAs) is positive from December to May and negative from July to September.

2. The correlation coefficient between the Niño3 SST deviation and the regional averaged ratio of annual precipitation to its long-term mean is under 0.55. In the warm phase of ENSO, there is a tendency that the annual precipitation anomalies are positive at eastern Japan facing the Pacific and western Japan from November to January, at the

[1] The data period is from 1979 to 2008.

DOI:10.1201/9781003603429-5

Ryukyu archipelago during December to March, and at western Japan from February to April.

Japan Meteorological Agency (JMA) has the criterion[2] for an El Niño to start, which is used to improve the accuracy of the medium- and long-range weather forecast.
In this chapter, the effect of five ENSO Indices (the SOI, the Niño1+2, Niño3, Niño3.4, Niño4 Indices)[3] on three meteorological parameters, the sea level pressure (SLP) anomalies (SLPAs), the surface air temperature ($\mathbb{SAT}$) anomalies ($\mathbb{SAT}$As), and the surface precipitation (SPR) anomalies (SPRAs), is investigated using the causality analysis.
Here, we will show the geography of Japan in Figure 5.1. Japan consists of four large islands (Hokkaido Island, Honshu Island, Shikoku Island, and Kyushu Island) and the Ryukyu archipelago (or the southwestern islands).

5.2 Data Used for This Study

Data used for this study are all monthly values, which are;

1. The Southern Oscillation Index (SOI)[4]

2. Four NiñoX (X is 1+2, 3, 3.4, 4) Indices (hereafter abbreviated NiñoXI) which are the normalized SST anomalies (SSTAs) of four Niño regions[5] [Reynolds et al., 2002].

3. The monthly $\mathbb{SAT}$, SLP, and SPR of 76 meteorological observatories in Japan[6].

The data period is from January 1982 to December 2019 (38 years or 456 months). The details of the SOI and the NiñoXI are described precisely in Chapter 4.
As for the $\mathbb{SAT}$, SLP, and SPR, we calculate the long-term (30 year) mean for each month (climatologies), which are subtracted from the actual data to make anomalies.

5.3 Effect of the SO on the Japanese Climate

In this section, we will discuss the cause-and-effect relationship between the SOI and three meteorological parameters using the causal analysis.

[2]The 5-month moving average of the SST departure of Niño3 region is over 0.5 °C warmer for successive 6 months or longer.

[3]As for the definition of the Indices of Niño four regions, refer Chapter 4.

[4]The SOI is downloaded from the website of National Centers for Environmental Information, National Oceanic and Atmospheric Administration (NOAA) (https://www.ncei.noaa.gov/access/monitoring/enso/soi).

[5]These data are provided by the PSL, NOAA, Boulder, Colorado, USA, from their website at https://psl.noaa.gov/data/timeseries/monthly/NINO12 etc.

[6]The data are provided by JMA through the website at https://www.data.jma.go.jp

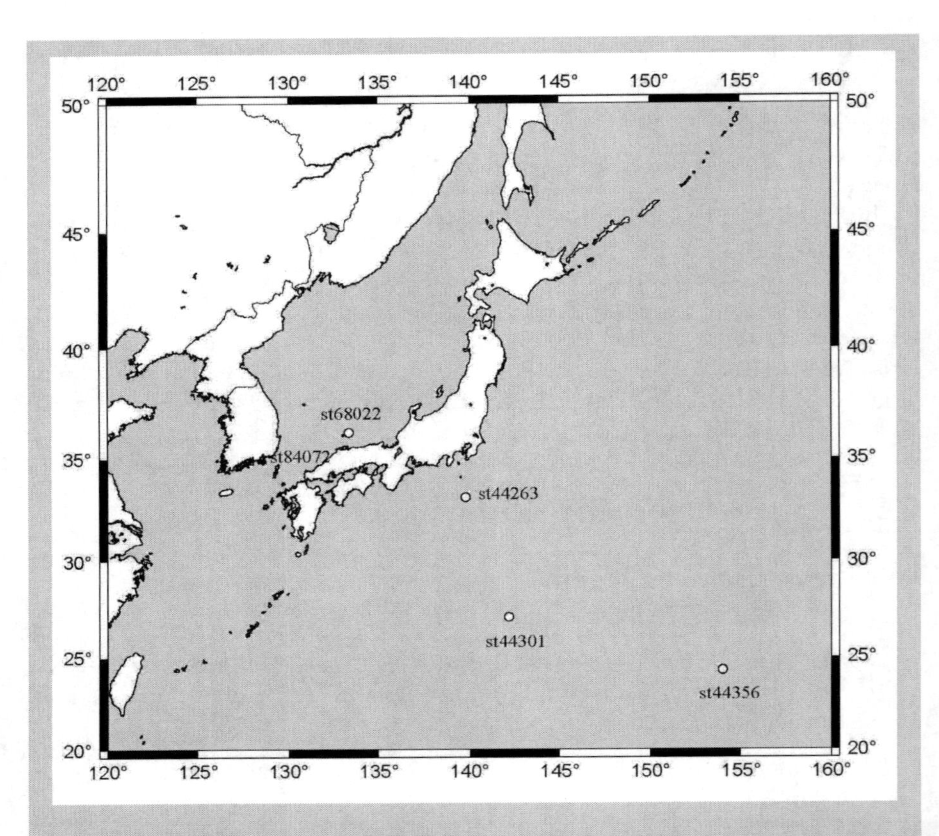

Figure 5.1: The geography of Japan. Marked observatories are Saigo (st68022), Iduhara (st84077), Hachijo-Jima (st44263), Chichi-Jima (st44301), and Minami-Torishima meteorological observatory (st44336).

5.3.1 Effect of the SOI on the SLPAs

At first, we compute the percentile of the causal value from the SOI to the SLPAs for 76 meteorological observatories. The spatial distribution of the percentile of the causal value from the SOI on the SLPAs is shown in Figure 5.2(a). The percentile of the causal value is higher than 95 at 75 stations, namely the SLPAs are intensely affected by the SOI all over Japan.
The only exception is the percentile of the Minami-Torishima meteorological observatory (st44356), which is lower than 50. We will consider the reason of this phenomenon. We expanded the SLP of the Minami-Torishima meteorological observatory to Fourier series, and the result is shown in Figure 5.3. The Fourier components are very close to those of the white noise. Referring Figure 4.2(b) which is the global structure of the percentile of the causal value from the SOI to the geopotential height anomalies at 1000 hPa level (GPHAs,

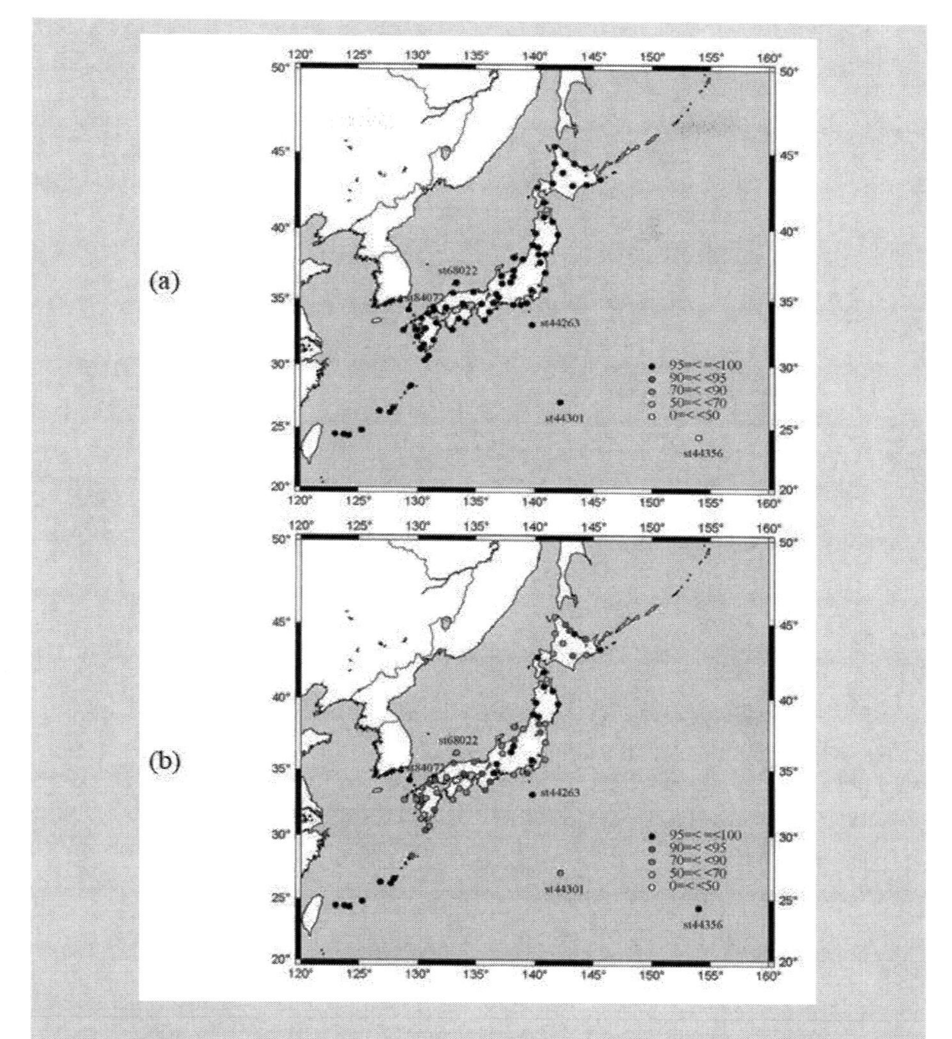

Figure 5.2: The spatial distribution of the percentile of the causal value from the SOI to the SLPAs (upper panel) and from the SOI to the SATAs (lower panel) of 76 meteorological observatories.

almost the same meteorological meaning as the SLPAs), we find the consistency that the Minami-Torishima station is included in the unaffected region extending from the tropical to subtropical central North Pacific, the mechanism of which is discussed in Chapter 4.

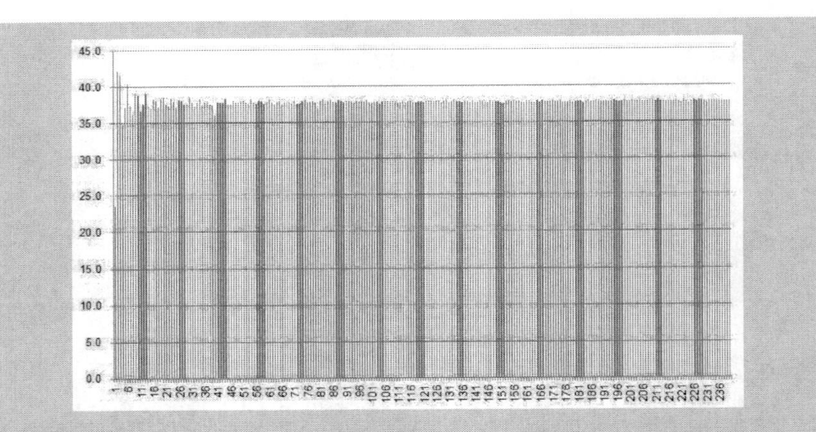

Figure 5.3: Fourier components of the SLP of the Minami-Torishima meteorological observatory (st44356). The abscissa is the period (unit is month) and the ordinate is the amplitude of each component (unit is hPa).

5.3.2 Effect of the SOI on the $\mathbb{SAT}$As

In this subsection, we compute the percentile of the causal value from the SOI to the $\mathbb{SAT}$As of the 76 meteorological observatories. The distribution of the percentile of the causal value is shown in Figure 5.3(b). The strongly affected meteorological observatories (the percentile is higher than 95) are in the Ryukyu archipelago (the southwestern islands), northern Honshu Island, southern Hokkaido Island, the Iduhara meteorological observatory (st84072) on a small island which locates at the Tsushima Channel between Kyushu Island and the Korean Peninsula, the Hachijo-Jima meteorological observatory (st44263), and the Minami-Torishima meteorological observatory (st44356).
The affected meteorological observatories (the percentile is between 90 and 95) exist at the Ryukyu archipelago, northern Kyushu Island, central and northern Honshu Island, and the northeastern coast of Hokkaido Island. Referring Figure 4.2 which is the global structure of the percentile of the causal value from the SOI to the air temperature anomalies at 1000 hPa level (SATAs, almost the same meteorological meaning as the $\mathbb{SAT}$As), the Ryukyu archipelago is the only affected region in Japan. The discrepancy between two results may be attributed to the difference of the data period used for the analysis.

5.3.3 Effect of the SOI on the SPRAs

In this subsection, we compute the percentile of the causal value from the SOI to the SPRAs of the 76 meteorological observatories. The distribution of the percentile of the causal value is not shown. The strongly affected meteorological observatories locate at western Kyushu Island, northern Hokkaido Island, and the Saigo meteorological observatory (st68022) on a small island in the western Japan Sea. The number of the strongly affected meteorological observatories is so few that it is difficult to find any meaningful spatial structure. However, they seem to align along the periphery of the Pacific High-pressure system.

5.4 Effect of the Niño1+2I on the Japanese Climate

In this section, we will discuss the cause-and-effect relationship between the Niño1+2I and three meteorological parameters using the causal analysis.

5.4.1 Effect of the Niño1+2I on the SLPAs

We compute the percentile of the causal value from the Niño1+2I to the SLPAs of 76 meteorological observatories. The distribution of the percentile of the causal value is shown in Figure 5.4(a). Five meteorological observatories are intensely affected by the Niño1+2I, which are three stations in Honshu Island, the Hachijo-Jima observatory (st44263), and the Chichi-Jima station (st44301). They seem to align about the periphery of the Pacific High on which the Niño1+2I can affect. Referring Figure 4.3(b) which is the global structure of the percentile of the causal value from the Niño1+2I to the GPHAs, southwestern Japan and the southern sea of Japan are strongly affected.

5.4.2 Effect of the Niño1+2I on the SATAs

In this subsection, we compute the percentile of the causal value from the Niño1+2I to the SATAs of the 76 meteorological observatories. The distribution of the percentile of the causal value is shown in Figure 5.4(b). The strongly affected meteorological observatories are nine, which are eight stations in the Ryukyu archipelago, and Minami-Torishima observatory (st44356). The possible mechanism is that the Niño1+2I affects the SATAs of these stations through the North Equatorial Current and the Kuroshio Current. Referring Figure 4.3(a) which is the global structure of the percentile of the causal value from the Niño1+2I to the SATAs, Japan is not affected at all. Both figures are inconsistent, but it may be attributed to the difference of the data period used for the analysis.

5.4.3 Effect of the Niño1+2I on the SPRAs

In this subsection, we compute the percentile of the causal value from the Niño1+2I to the SPRAs of the 76 meteorological observatories. The SPRA of all observatories are not affected by the Niño1+2I.

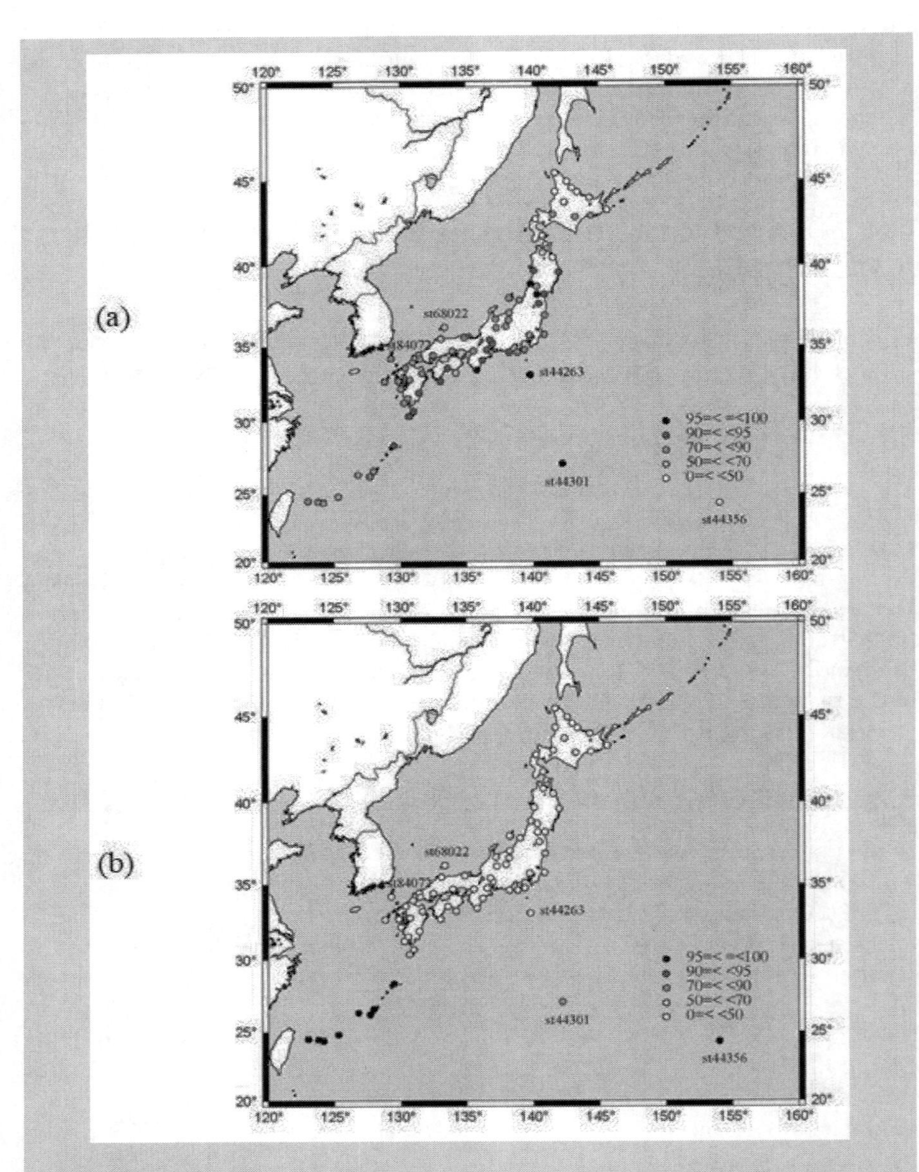

Figure 5.4: The spatial distribution of the percentile of the causal value from the Niño1+2I to the SLPAs (upper panel) and from the Niño1+2I to the SATAs (lower panel) of 76 meteorological observatories.

5.5 Effect of the Niño3I on the Japanese Climate

In this section, we will discuss the cause-and-effect relationship between the Niño3I and three meteorological parameters using the causal analysis.

5.5.1 Effect of the Niño3I on the SLPAs

We compute the percentile of the causal value from the Niño3I to the SLPAs of 76 meteorological observatories. The distribution of the percentile is shown in Figure 5.5(a). Thirteen meteorological observatories are strongly affected by the Niño3I, eleven stations in Honshu Island, the Hachijo-Jima observatory (st44263), and the Chichi-Jima meteorological observatory (st44301). Meteorological observatories whose percentile of the causal value is between 90 and 95 distribute near the strongly affected observatories. Their distribution seems to coincide with the periphery of the Pacific High-pressure system on which the Niño3I can affect. Taking into account the difference of the data period, we will compare Figure 5.5(a) with Figure 4.4(b) which is the global structure of the percentile of the causal value from the Niño3I to the GPHAs. We find the consistency between two figures, namely the area including the Chichi-Jima meteorological observatory is intensely affected but the area including the Minami-Torishima meteorological observatory is not affected by the Niño3I.

5.5.2 Effect of the Niño3I on the $\mathbb{SAT}$As

In this subsection, we compute the percentile of the causal value from the Niño3I to the $\mathbb{SAT}$As of the 76 meteorological observatories. The distribution of the percentile of the causal value is shown in Figure 5.5(b). The strongly affected meteorological observatories are seven stations at the Ryukyu archipelago, and the Chichi-Jima meteorological observatory (st44301). The $\mathbb{SAT}$As of the Ryukyu archipelago and the Chichi-Jima Island may be affected by the Niño3I through the North Equatorial Current and the Kuroshio Current, which often meanders southward departing the southern coast of Japan at the offshore of eastern Shikoku Island, passes near Chichi-Jima Island, and returns near the cost of Japan at the Hachijo-Jima observatory (st44263). Referring Fig. 4.4(a) which is the global structure of the percentile of the causal value from the Niño3I on the SATA, the SATAs of Japan is not affected by the Niño3I.

5.5.3 Effect of the Niño3I on the SPRAs

In this subsection, we compute the percentile of the causal value from the Niño3I to the SPRAs of the 76 meteorological observatories. The distribution of the percentile of the causal value is not shown. The affected meteorological

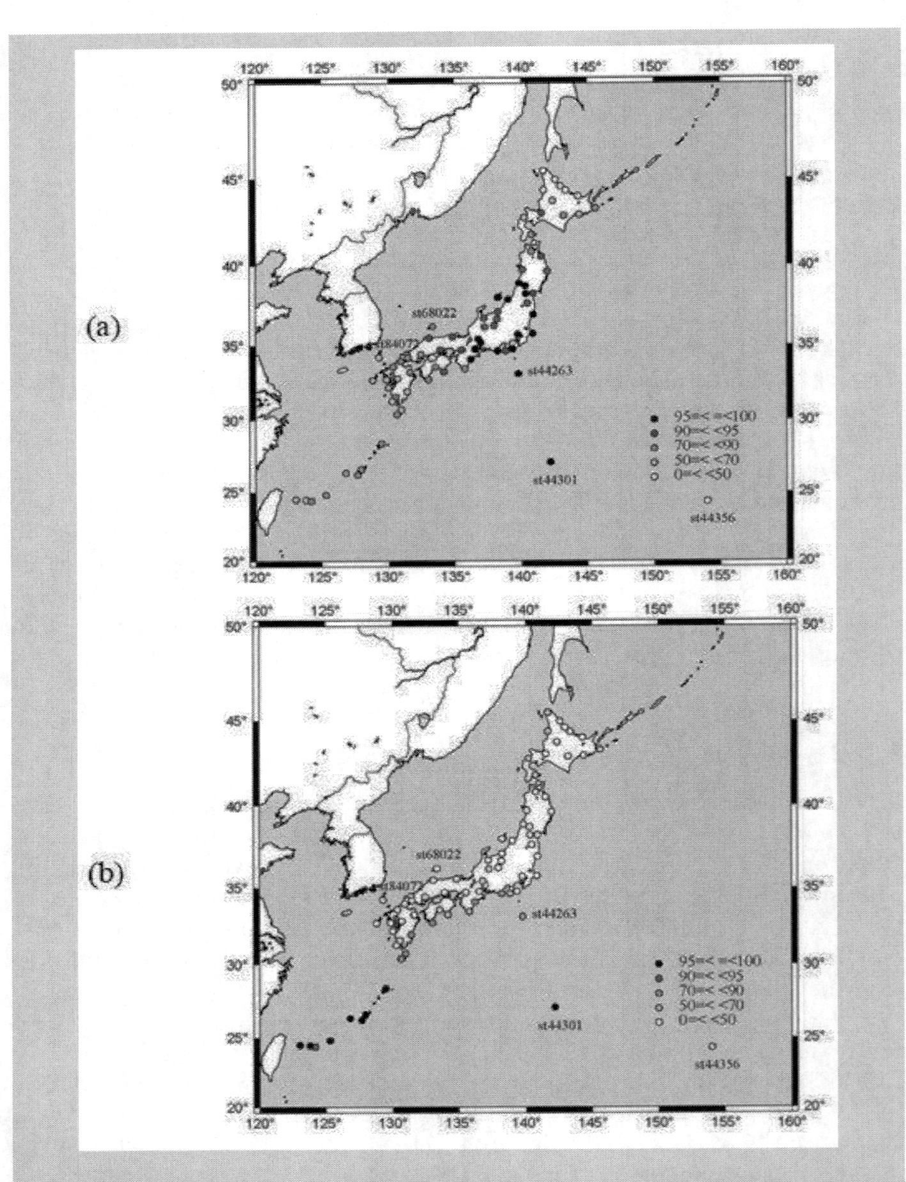

Figure 5.5: The spatial distribution of the percentile of the causal value from the Niño3I to the SLPAs (upper panel) and from the Niño3I to the SATAs (lower panel) of 76 meteorological observatories.

observatories are only three, two are in the Ryukyu archipelago and one is at northern Kyushu Island.

5.6 Effect of the Niño3.4I on the Japanese Climate

In this section, we will discuss the cause-and-effect relationship between the Niño3.4I and three meteorological parameters of 76 meteorological observatories using the causal analysis.

5.6.1 Effect of the Niño3.4I on the SLPA

We compute the percentile of the causal value from the Niño3.4I to the SLPAs of 76 meteorological observatories. The distribution of the percentile of the causal value is shown in Figure 5.12. The intensely affected observatories form the north-south band structure, which seems to be the periphery of the Pacific High-pressure system on which the Niño3.4I can affect. So, we guess that the rise and fall of the SLP of the Pacific High is strongly affected by the Niño3.4I. Referring Figure 4.32 which is the global structure of the percentile of the causal value from the Niño3.4I to the GPHAs, the GPHAs of Japan except Hokkaido Island and the Minami-Torishima meteorological observatory (st44356) is strongly affected. Both results are consistent except Kyushu Island.

5.6.2 Effect of the Niño3.4 Index on the $\mathbb{SAT}$As

In this subsection, we compute the percentile of the causal value from the Niño3.4I to the $\mathbb{SAT}$As of the 76 meteorological observatories. The distribution of the percentile of the causal value is shown in Figure 5.6(b), which is almost similar to Figure 5.5(b). The affected meteorological observatories are seven stations at the Ryukyu archipelago and the Chichi-Jima meteorological observatory (st44301). The $\mathbb{SAT}$As of these meteorological observatories may be affected by the Niño3.4I through the North Equatorial Current and the Kuroshio Current. Referring Figure 4.5(a) which is the global structure of the percentile of the causal value from the Niño3.4I to the SATAs, the affected observatories in Figure 5.6(b) locate at the strongly affected region in Figure 4.5(a).

5.6.3 Effect of the Niño3.4I on the SPRAs

In this subsection, we compute the percentile of the causal value from the Niño3.4I to the SPRAs of the 76 meteorological observatories. The distribution of the percentile of the causal value is not shown. The SPRA of only five meteorological observatories are affected, which are the west most station of

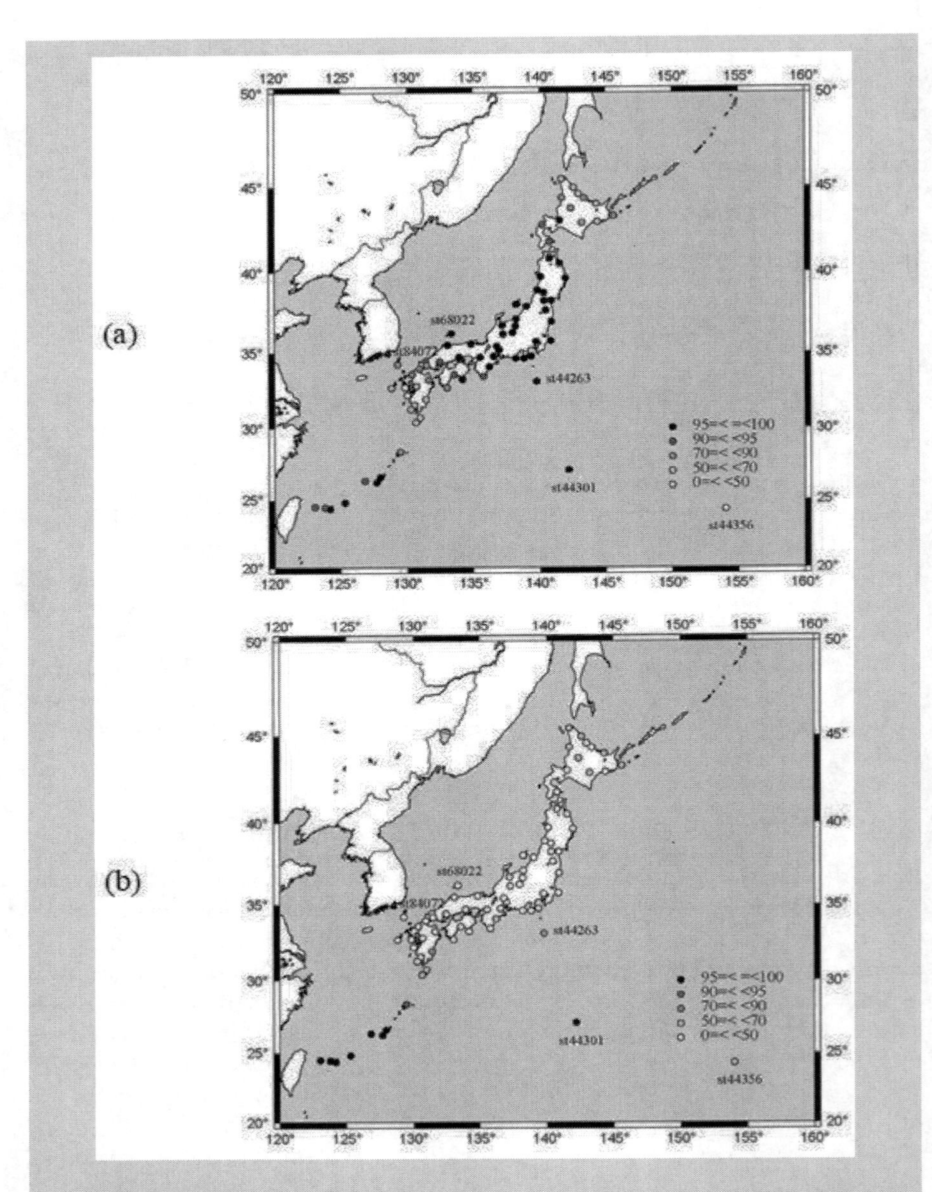

Figure 5.6: The spatial distribution of the percentile of the causal value from the Niño3.4I to the SLPAs (upper panel) and from the Niño3.4I to the SATAs (lower panel) of 76 meteorological observatories.

Honshu Island, the Saigo meteorological observatory (st68022), two observatories of central Honshu Island, and the south most observatory of Hokkaido Island. The affected observatories are so few and sparce that we cannot find out the meaningful spatial structure.

5.7 Effect of the Niño4I on the Japanese Climate

In this section, we will discuss the cause-and-effect relationship between the Niño4I and three meteorological parameters using the causal analysis.

5.7.1 Effect of the Niño4I on the SLPAs

We compute the percentile of the causal value from the Niño4I to the SLPAs of 76 meteorological observatories. The distribution of the percentile is shown in Figure 5.7(a), which resembles Figure 5.6(a). Eight meteorological observatories in the Ryukyu archipelago, seventeen stations in Honshu Island, one station in Hokkaido Island, and the Chichi-Jima meteorological observatory (st44301) are intensely affected. These meteorological observatories seem to align about the periphery of the Pacific High on which the Niño4I is able to affect. Referring Figure 4.6(b) which is the global structure of the percentile of the causal value from the Niño4I to the GPHAs, southwestern Japan is strongly affected.

5.7.2 Effect of the Niño4I on the $\mathbb{SAT}$As

In this subsection, we compute the percentile of the causal value from the Niño4I to the $\mathbb{SAT}$As of the 76 meteorological observatories. Six affected meteorological observatories all locate at the Ryukyu archipelago (a figure is not shown). This region may be affected by the Niño4I through the North Equatorial Current and the Kuroshio Current. Referring Figure 4.6(a) which is the global structure of the percentile of the causal value from the Niño4I to the SATAs, the SATAs of Japan except the Ryukyu archipelago is not affected by the Niño4I.

5.7.3 Effect of the Niño4I on the SPRAs

In this subsection, we compute the percentile of the causal value from the Niño4I to the SPRAs of the 76 meteorological observatories. The distribution of the percentile of the causal value is shown in Figure 5.7(b). The affected meteorological observatories are only three, which are the west most station of Honshu Island, an observatory at northern Kyushu Island and the observatory at the southern small island of Kyushu Island. It is difficult to find out some

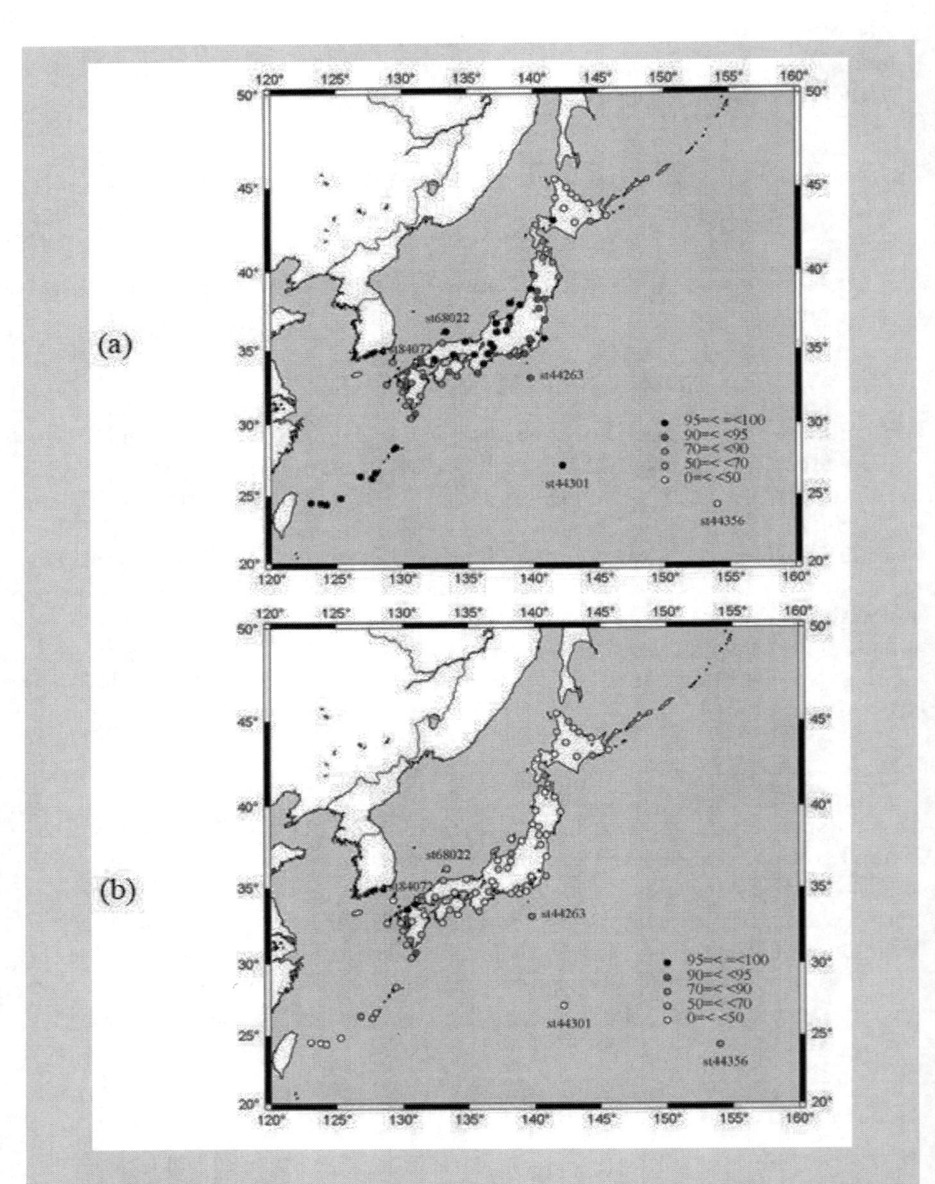

Figure 5.7: The spatial distribution of the percentile of the causal value from the Niño4I to the SLPAs (upper panel) and of from the Niño4I to the SPRAs (lower panel) of 76 meteorological observatories.

systematic spatial structure, but they seem to locate at the periphery of the Pacific High on which the Niño4I can affect.

5.8 Summary and Discussions

It is proved mathematically that the ENSO phenomena influence seriously the Japanese climate. As for the SLPAs, the SOI affects most meteorological observatories of Japan, namely 75 observatories out of 76. One exception is Minami-Torishima meteorological observatory (st44356), which locates at the subtropical central North Pacific. Four Niño regions all affect the SLPAs of Japan. The affected meteorological observatories form the north-south lines or bands, which are supposed to be the periphery of the Pacific High-pressure system on which the NiñoXI can affect. The north-south lines or bands shift westward according to the distance between Japan and the NiñoX regions (Figures 5.4(a), 5.5(a), 5.6(a), and 5.6(a)). The number of stations affected by the NiñoXI increases as the regions shift nearer to Japan.
It is worth to compare Figure 5.2(a) with Figures 4.2(b) and 4.7, which are the distribution of the percentile of the causal value from the SOI to the GPHAs. We should take care that the data period is different in Figures 4.2(b) and 4.7, namely 144 months (12 years) in the former and 456 months(38 years) in the latter. In Figure 4.2(b), the SOI affects the GPHAs in southern Japan. While in Figure4.7, the GPHAs of whole Japan is intensely affected by the SOI. In both figures, there is an unaffected region in the tropical and subtropical central North Pacific, where the Minami-Torishima meteorological observatory (st44356) exists.
We will consider next about the effect of five ENSO Indices on the SATAs of Japan. The SOI strongly affects the Minami-Torishima meteorological observatory, the Chichi-Jima meteorological observatory (st44301), five stations in the Ryukyu archipelago, and fourteen observatories of Kyushu, Honshu, and Hokkaido Island, and it affects eighteen stations all over the country (Figure 5.2(b)). The Niño1+2I affects the SATAs of seven meteorological observatories of the Ryukyu archipelago (Figure 5.4(b)), the Niño3I affects the Chichi-Jima meteorological observatory and eight stations in the Ryukyu archipelago (Figure 5.5(b)), the Niño3.4I affects the same observatories as those of the Niño3I (Figure 5.6(b)), and the Niño4I affects meteorological observatories almost the same as those of the Niño3I and the Niño4I except Chichi-Jima meteorological observatory. The distributions of the affected meteorological observatories by four NiñoXI are almost similar, which is the reflection that the SATAs of Japan is strongly affected by the SSTAs of the NiñoX regions through the North Equatorial Current and the Kuroshio Current.
Finally, we will discuss about the effect of five ENSO Indices on the SPRAs of Japan. The SOI affects several observatories in Kyushu Island, one in Shikoku Island, three stations in Honshu Island, and four observatories in Hokkaido Island, which seem to locate the periphery of the Pacific High-pressure system.

The Niño1+2I does not affect the SPRAs of Japan, the Niño3I affects two meteorological observatories of the Ryukyu archipelago, and one in Kyushu Island, the Niño3.4I affects only four stations, from which we cannot find any systematic structure. The Niño4I affects ten stations in Kyushu Island and one in Honshu Island (Figure 5.7(b)). Generally speaking, the influence of the NiñoXI on the SPRAs of Japan is very weak and non-systematic. This tendency is also found for the global structure of the percentile of the causal value from the NiñoXI on the SPRAs of the 144×72 grid points (refer Chapter 4).

Chapter 6

The Effect of the NAO on the Global Climate

by Osamu Morita

In this chapter, we will discuss the effect of the North Atlantic Oscillation on the global climate parameters, the air temperature at the 1000 hPa level, the geopotential height of the 1000 hPa surface, and the surface precipitation, using the theory of the causal analysis.

6.1 Introduction

The North Atlantic Oscillation (NAO) is one of the major atmospheric variability in the Northern Hemisphere [Hurrell, 1995]. It controls the strength and direction of westerlies, and the route of storm tracks, so that it affects the surface air temperature ($mathbbSAT$) and the surface precipitation (SPR) of the northern North America, Greenland, Europe, and far to Siberia. The effect of the NAO is especially strong in the boreal winter. The NAO Index (NAOI) is defined as the measure of the strength of the NAO, whose historical definition is the difference between the normalized sea level pressure (SLP) of the stations in western Spain which represent the SLP of the Azores High and the normalized SLP of the station at southwestern Iceland which represents the SLP of the Icelandic Low[1]. In the highly positive phase of the NAOI, enhanced westerlies bring about cool summers and mild and wet winters to northern and central Europe. In the negative phase of the NAOI, northern Europe faces hot summers and cold and dry winters due to suppressed westerlies.

[1] There are a few combinations of stations: (1) Lisbon and Stykkishóllmur/Reykjavík (2) Ponta Delgada, Azores and Stykkishólmur/Reykjavík (3) Azores, Gibraltar and Reykjavík.

DOI:10.1201/9781003603429-6

We will discuss the effect of the NAO on the global climate in the following sections. The data used in this study are introduced in section 6.2, the causal relation between the NAO, SO, and Arctic Oscillation (AO) is discussed in section 6.3, the effect of the NAO on the global climate is followed in section 6.4, and discussion and summary are developed in section 6.5.

6.2 Data Used for This Study

Data used for this study are all monthly values, which are;

1. The North Atlantic Oscillation (NAO) index (NAOI)[2].

2. The Southern Oscillation (SO) Index (SOI)[3]

3. The monthly air temperature at the 1000 hPa level (SAT), and the monthly geopotential height of the 1000 hPa surface (GPH) of the NCEP/NCAR Reanalysis 1 data[4][Kalnay et al., 1996].

4. The monthly surface precipitation (SPR) of CMAP (CPC[5] Merged Analysis of Precipitation) data[6] [Xie and Arkin, 1997; Schneider et al., 2013].

The NAOI is defined by the difference between the normalized SLP of a station under the influence of the Azores High and a station under the influence of the Icelandic Low [Hurrell, 1995; Jones et al., 1997]. As for the precise calculation of the NAOI, refer the article of the home page of Climate Research Unit, University of East Anglia.
The SAT, and the GPH are the global gridded data at 144×73 points with the data period from January 1981 to December 2010. We calculate the long-term (30 year) mean for each month (climatologies), which are subtracted from the actual data to make anomalies. The anomalies of the SAT and the GPH are designated SATAs and GPHAs. Finally, we use the 240-month (from January 1991 to December 2010) data for this study.
The SPR are the global gridded data at 144×72 points with the data period from January 1981 to December 2010. We made the time series of the SPR anomalies (SPRAs) as the same procedure mentioned above. We use the 240-month (from January 1991 to December 2010) data for this study.

[2]The NAOI is downloaded from the website of Climate Research Unit, University of East Anglia (https://crudata.uae.ac.uk/data/nao/).

[3]The SOI is downloaded from the website of National Centers for Environmental Information, National Oceanic and Atmospheric Administration (NOAA) (https://www.ncei.noaa.gov/access/monitoring/enso/soi).

[4]These data are provided by the PSL, NOAA, Boulder, Colorado, USA, from their website at https://psl.noaa.gov

[5]Climate Prediction Center, NOAA

[6]These data are provided by the PSL, NOAA, Boulder, Colorado, USA, from their website at https://psl.noaa.gov/data/gridded/data.cmap.html

6.3 The Causal Relation between the NAO and the SO

As for the relationship between the NAO and the ENSO, there are many studies so far. It is known that the NAOI and the ENSO Index reach synchronously their peak in the boreal winter. However, Hurrell [1995] showed that there is no correlation between the NAOI and the Nino3.4 Index of their winter time mean (the correlation coefficient is 0.06). Freadrich and Müller [1992] found that in the northern Scandinavian Peninsula the SLPAs are positive, the SATAs are negative, and the SPRAs are negative in the EL Niño phase, and vice versa in the La Niña phase. Huang et al. [1998], and Cassou and Terray [2001] obtained the similar result as Freadrich and Müller [1992]. Quadrelli and Wallace [2002] showed that the structure of the AO changes drastically due to the phase of ENSO. Bronninmann et al. [2004] analyzed the strong El Niño from 1940 to 1942, and found that the Aleutian Low-pressure system was intensified and the NAOI was in the negative phase in the winter season in the troposphere, and the circum-polar vortex was weak in the lower stratosphere. Here, we will study the causal relationship between the NAO and the SO, between the NAO and the AO, and between the SO and the AO. In the calculation of the percentile of the causal value, we examine the dependence of the percentile on the data number.
In Table 6.1, we show the data number dependence of the percentile of the causal value between the NAOI and the SOI. The percentile becomes stable or saturated when the data number exceeds 210 months. Taking into account this result, we have chosen the data number 240 months in this chapter. The minimum significant data number seems to depend on the signal-to-noise ratio, the period of the objective phenomenon, and so on.
The NAO affects intensely the SO but the SO does not affect the NAO at all. There is no causal relationship between the SO and the AO, and between the NAO and the AO (the results are not shown).

Table 6.1: The percentile of the causal value between the NAOI and the SOI, showing the dependence of the percentile of the causal value on the data number.

month	from SOI	from NAOI
120	58.6	89.6
144	41.2	90.3
180	7.0	85.7
210	29.4	92.5
240	28.9	96.6
300	25.3	95.7
360	19.7	96.2

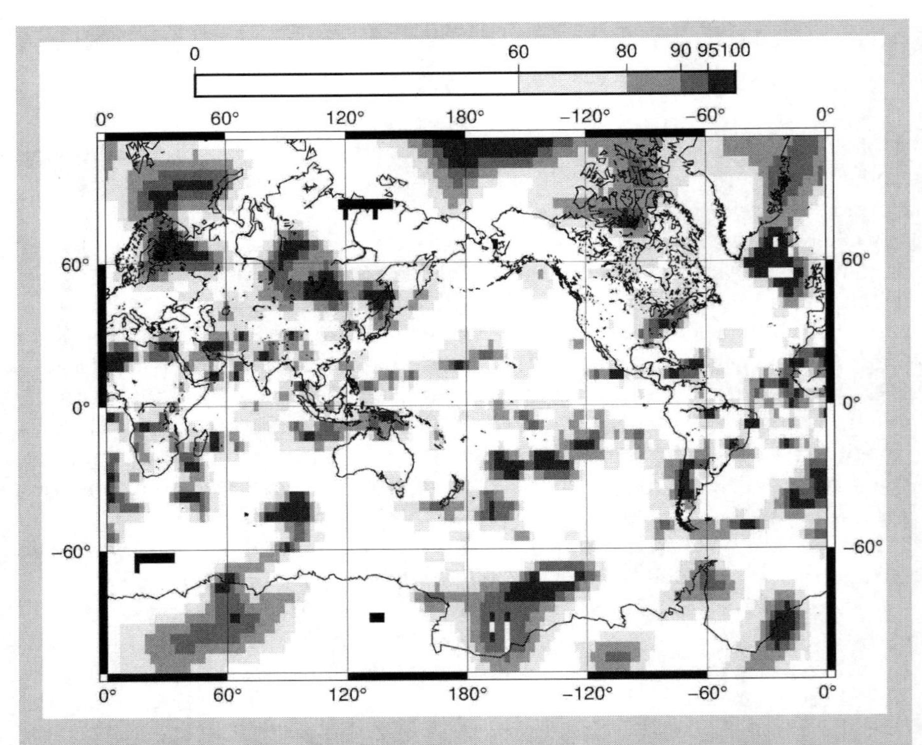

Figure 6.1: The spatial structure of the percentile of the causal value from the NAO to the SATAs on the conformal Mercator projection (80°N–80°S) centering at 0°, 180°E.

6.4 The Effect of the NAO on the Global Climate Elements

In this section, we will discuss the causal effect of the NAO on the three global climate parameters, SATAs, GPHAs, and SPRAs.

6.4.1 The Effect of the NAO on the Global SATAs

In Figure 6.1, we show the spatial structure of the percentile of the causal value from the NAO to the SATAs on the global 144×73 grid points. Figure 6.1 is the conformal Mercator projection from 80°N to 80°S centering at 180°E and 0°.

In the northern high latitudes and within the Arctic Circle, we find that the NAO affects the SATAs at the northeastern North Atlantic, the east coast of Greenland, the Arctic Sea centering at 180°E, central Russia facing the Arctic

Sea, northern Europe, southern Fennoscandia, from the Barents Sea to the western sea of Novaya Zemlya Islands, Russia.
In the northern midlatitudes, the NAO affects the SATAs at eastern United States, the northern Japan Sea, and the central Eurasian Continent.
In the tropics, the affected regions are the tropical Atlantic Ocean, the Caribbean see, New Caledonia, the western sea of Australia, the sea surrounding Madagascar, and tropical Africa.
In the southern midlatitudes, the SATAs is affected in the southern South Atlantic, the central South Pacific, the southern Indian Ocean, and the southeastern and southwestern sea of South Africa .
In the southern high latitudes and around Antarctica, the intensely affected regions are southern Chile, the Antarctic Peninsula, the Ross Sea, the offshore sea of Queen Mary Land, MacRobertson Land, and Dronning Maud Land.
It is the common recognition that the NAO affects the Northern Hemispheric climate [Hurrell, 1995; Hurrell et al., 2003; Osborn, 2011]. However, the NAO influences the Southern Hemispheric climate in the present study. We show in section 6.2 that the NAO has the intense causal effect on the SO, so that the NAO may influence the Southern Hemispheric climate through the SO.

6.4.2 The Effect of the NAO on the Global GPHAs

In Figure 6.2, we show the spatial structure of the percentile of the causal value from the NAO to the GPHAs on the global 144×73 grid points. The figure is the conformal Mercator projection from 80°N to 80°S centering at 180°E and 0°.
In the northern high latitudes and within the Arctic Circle, we find that the NAO affects the GPHAs at southern Fennoscandia, northeastern Europe, and western Russia.
In the northern midlatitudes, the NAO affects the GPHAs at the central North Atlantic Ocean, central and eastern United States, southwestern Japan, and southwestern Russia.
In the tropics, the affected regions are the tropical South Pacific, northern Australia, the tropical Indian Ocean, the Red Sea, the southern sea of the Arabian Peninsula, Madagascar, and western Angola.
In the southern midlatitudes, the affected regions are the southern South Atlantic, central Argentine, central Chile, and the central South Pacific.
In the southern high latitudes and around Antarctica, the affected regions are the King Haakon VII Sea, the Bellingshausen Sea, and the Amundsen Sea.

6.4.3 The Effect of the NAO on the Global SPRAs

In Figure 6.3, we show the spatial structure of the percentile of causal values from the NAO to the SPRAs on the global 144×72 grid points. The figure is the conformal Mercator projection from 80°N to 80°S centering at 180°E and 0°.

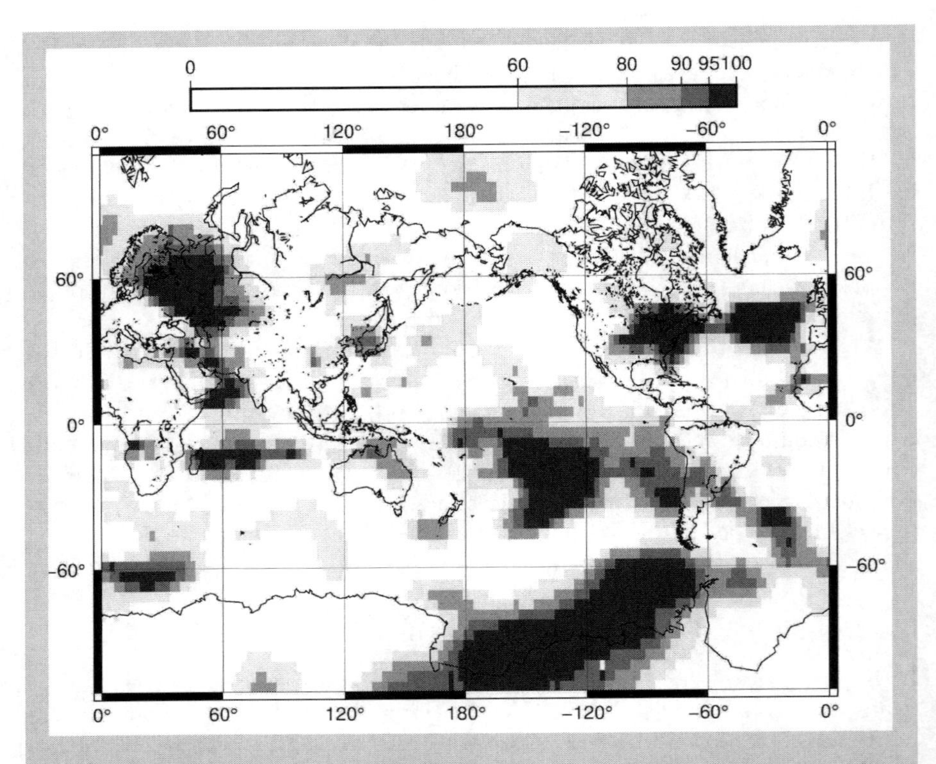

Figure 6.2: The spatial structure of the percentile of the causal value from the NAO to the GPHAs on the conformal Mercator projection (80°N–80°S) centering at 0°, 180°E.

In the northern high latitudes and within the Arctic Circle, we find that the NAO affects the SPRAs at southern Greenland, eastern Alaska of United States, and the northwestern sea of Norway.
In the northern midlatitudes, the NAO affects the SPRAs at the central North Atlantic, and the small region north of the Caspian Sea.
In the tropics, the affected regions are the tropical Atlantic Ocean, the western tropical Pacific, the tropical Indian Ocean, and western Angola.
In the southern midlatitudes, the affected regions are southern Argentine, southern Chile, the southeastern South Pacific, eastern Australia, and southern South Africa.
In the southern high latitudes and around Antarctica, the affected regions are the most part of West Antarctica, Princess Elizabeth Land, Wilhelm II Land, and Queen Mary Land.
In general, the effect of the NAO on the SPRAs is rather weak, and the affected regions are small and sparce.

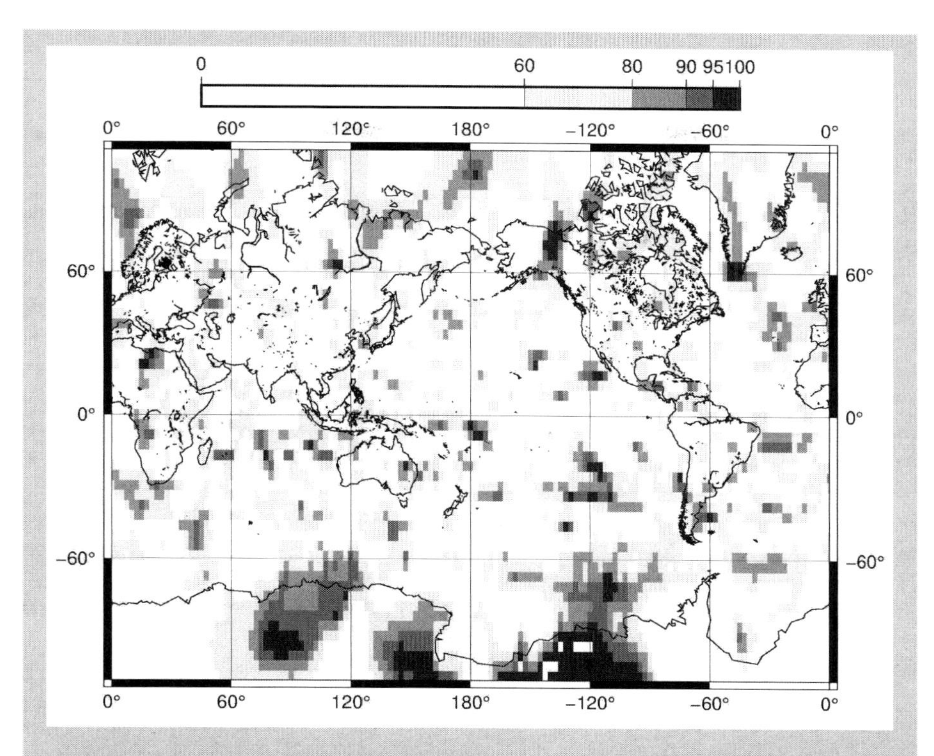

Figure 6.3: The spatial structure of the percentile of the causal value from the NAO to the SPRAs on the conformal Mercator projection (80°N–80°S) centering at 0°, 180°E.

6.5 Discussions and Summary

Jones et al. [2003] investigated the correlation between the NAOI for the winter season (December to February) and the SAT and SPR for two periods 1901–1950 and 1951–2000. Their findings are:

1. The relationship between the NAOI and SAT is strongest over southern Fennoscandia.

2. The relationship is inverse south of 40°N, particularly over the southern Balkans, Turkey, parts of the Middle East, southern Spain, and north-western Africa.

3. All the main centers of influence over the North Atlantic, Europe, and eastern Noth America are more coherent and stronger during 1951–2000 than 1901–1950.

4. The precipitation influence extends over a much smaller region, being limited to northern Europe, the Mediterranean, and North Africa.

5. Correlation with precipitation are also weaker during the 1901–1950 period.

Referring Figure 6.1, the percentile of the causal value from the NAOI to the SATAs is higher than 95 at southern Fennoscandia, norther Europe, the northern North Atlantic, which is consistent with the result of Jone et al. [2003]. As for the SPR, the affected regions are small and sparce (Figure 6.3), whose tendency coincides with their result. In general, the response of the SPR to teleconnection indices (e.g. the NAO, SOI, and NiñoX Index) is weaker than that of the SAT and GPH (refer Chapter 4). This fact may come from that the SPR is affected by many meteorological parameters, namely the air temperature, the distribution of the low- and high-pressure system, the distribution of water vapor, the path of midlatitude and tropical depressions, and so on.
It is a very important finding that the NAO influences strongly the SO. The possible mechanism how the NAO affects the SO is found in the papers of Hurrell [1995] and Hurrell et al. [2003]. They say that the NAO affects not only the peripheral region of the North Atlantic Ocean but also the Eurasian continent and the North Pacific. The SLPA of the area around Iceland and the area of the northern North Pacific are out of phase, so that the NAO influences the tropospheric air circulation all over the Northern Hemisphere. Through the process above mentioned there is a possibility that the NAO influences the westerly surge at tropics and be a trigger of the El Niño. Referencing Figure 6.4, we can guess another possible mechanism that the NAO influences the ENSO. The SLPA of the Azores High-pressure system affects directly the SLPA of the tropical eastern Pacific, which could be a trigger of the El Niño.

Appendix A

Abbreviations

Abbreviations used in this book are listed as follows.

AO	Arctic Oscillation
AR	autoregressive
ARMA	autoregressive moving average
CMAP	CPC merged analysis of precipitation
CPC	Climate Prediction Center, NOAA
CRU	Climatic Research Unit, University of East Anglia
DVI	dust veil index
ENSO	El Niño and Southern Oscillation
FDT	Fluctuation–Dissipation Theorem
GPH	geopotential height at 1000 hPa level
GPHA	geopotential height anomalies at 1000 hPa level
JMA	Japan Meteorological Agency
KM_2O	Kubo, Mori, Miyoshi, and Okabe
NAO	North Atlantic Oscillation
NAOI	North Atlantic Oscillation Index
NASA	National Aeronautics and Space Administration
NCAR	National Center for Atmospheric Research
NCEP	National Centers for Environmental Prediction
NOAA	National Oceanic and Atmospheric Administration
PSL	Physical Sciences Laboratory, NOAA
QBO	quasi-biennial oscillation
RSAT	regional averaged surface air temperature
SAA	sulfuric acid aerosol
SAGE Ⅱ	stratospheric aerosol and gas experiment Ⅱ
SAT	air temperature at 1000 hPa level
SATA	air temperature anomaly at 1000 hPa level
$\mathbb{SAT}$	surface air temperature
$\mathbb{SATA}$	surface air temperature anomaly

DOI:10.1201/9781003603429-A

SEA	superposed epoch analysis
SLP	sea level pressure
SLPA	sea level pressure anomaly
SO	Southern Oscillation
SOI	Southern Oscillation Index
SPR	surface precipitation
SPRA	surface precipitation anomaly
SST	sea surface temperature
SSTA	sea surface temperature anomalies

Appendix B

Programming Codes for Causal Analysis

All programming codes are written in "True Basic" invented by J.G. Kemeny and T.E. Kurtz, and are free to use to confirm our results and/or to be applied for your studies. The programming codes are also permitted to convert to other programming languages.

B.1 Calculation of the Causal Value

The programming code is the algorism for calculating the causal value from the time sequential data $\mathcal{Y}$ to the time sequential data $\mathcal{X}$. The causal value is stored in the file "Cval.dat". The programming code is as follows:

Programming Code

```
! Program Name: caxy.tru: Test of Stationarity
!This program was made by Yuji Nakano
DIM A1(2,2),B1(2,2),C1(2,2),D1(2,2),F1(2,2),G(2,2),E(2,2)
DIM A2(2,2),B2(2,2),C2(2,2),D2(2,2),F2(2,2)
DIM T(2,2),TINV(2,2),H(2,1),TH(2,1)
!
OPTION BASE 0
!
DIM X1(1000),X2(1000)
DIM X(2,1000),Z(2,1000),Z1(2,100),R(2,2,100),TR(2,2,100)
DIM DE1(100),GA1(100,100),V1(100)
DIM DEP(2,2,100),DEM(2,2,100),VP(2,2,100),VM(2,2,100)
DIM GAP(2,2,100,100),GAM(2,2,100,100)
DIM FPZ(2,100),WP(2,100),WZ(100)
!
PRINT "Dimension of data: D=2"
```

DOI:10.1201/9781003603429-B

```
LET ID=2
!
INPUT PROMPT "data number?": N
INPUT PROMPT"1-st file name to read as ******(.dat)":T1$
INPUT PROMPT"2-nd file name to read as ******(.dat)":T2$
LET N=N−1
LET T1$=T1$ & ".dat"
LET T2$=T2$ & ".dat"
LET T3$="Cval.dat"
!
! Reading data from data files
OPEN #1: NAME T1$, ORG TEXT, CREATE OLD, ACCESS INPUT
FOR I=0 TO N
INPUT #1: X1(I)
NEXT I
CLOSE #1
!
OPEN #2: NAME T2$, ORG TEXT, CREATE OLD, ACCESS INPUT
FOR I=0 TO N
INPUT #2: X2(I)
NEXT I
CLOSE #2
FOR I=0 TO N
LET X(1,I)=X1(I)
LET X(2,I)=X2(I)
NEXT I
FOR J=1 TO ID ! Minus mean
LET S=0
FOR I=0 TO N
LET S=S+X(J,I)
NEXT I
LET XBAR=S/(N+1)
FOR I=0 TO N
LET X(J,I)=X(J,I)−XBAR
NEXT I
NEXT J
!
FOR J=1 TO ID      ! Divided by standard deviation
LET SS=0
FOR I=0 TO N
LET SS=SS+X(J,I)*X(J,I)
NEXT I
LET XVAR=SS/(N+1)
FOR I=0 TO N
LET Z(J,I)=X(J,I)/SQR(XVAR)
```

```
NEXT I
NEXT J
!
!****************************************************************
LET M1=INT(3*SQR(N+1))-1
LET MD=INT(2*SQR(N+1)/ID)-1      !Efficient number of R
LET MM=MD
!****************************************************************
! Covariance of normalized data
FOR J1=1 TO ID
FOR J2=1 TO ID
FOR J=0 TO M1
LET CV=0
FOR I=0 TO N-J
LET CV=CV+Z(J1,I+J)*Z(J2,I)
NEXT I
LET R(J1,J2,J)=CV/(N+1) ! Covariance Function
NEXT J
NEXT J2
NEXT J1
!
!************** ALGORITHM : 1-DIM *************************
!********************************************************
! Algorithm of KM2O-Langevin equation
!********************************************************
LET DE1(1)=-R(1,1,1)/R(1,1,0) ! 1st STEP
LET GA1(1,0)=DE1(1)
LET V1(0)=R(1,1,0)
LET V1(1)=(1-DE1(1)*DE1(1))*V1(0)
!
LET DE1(2)=-(R(1,1,1)*DE1(1)+R(1,1,2))/V1(1) ! 2nd STEP
LET GA1(2,0)=DE1(2)
LET GA1(2,1)=DE1(1)*(1+DE1(2))
LET V1(2)=(1-DE1(2)*DE1(2))*V1(1)
!
FOR I=3 TO M1 ! General Step
LET S1=0
FOR J=0 TO I-2
LET S1=S1+GA1(I-1,J)*R(1,1,J+1)
NEXT J
LET DE1(I)=-(R(1,1,I)+S1)/V1(I-1)
LET V1(I)=(1-DE1(I)*DE1(I))*V1(I-1)
LET GA1(I,0)=DE1(I)
FOR J=1 TO I-1
LET GA1(I,J)=GA1(I-1,J-1)+DE1(I)*GA1(I-1,I-1-J)
```

```
NEXT J
NEXT I
!
!************** ALGORITHS : 2-DIM ****************************
!1st-Step
FOR J1=1 TO ID
FOR J2=1 TO ID
LET G(J1,J2)=R(J1,J2,0)
NEXT J2
NEXT J1
!
FOR J1=1 TO ID
FOR J2=1 TO ID
LET VP(J1,J2,0)=G(J1,J2)
LET VM(J1,J2,0)=G(J1,J2)
NEXT J2
NEXT J1
!
MAT A1=INV(G) !Inverse of R(0)
!
FOR J1=1 TO ID !R(1)
FOR J2=1 TO ID
LET B1(J1,J2)=−R(J1,J2,1)
LET B2(J1,J2)=−R(J2,J1,1)
NEXT J2
NEXT J1
!
MAT C1=B1*A1
MAT C2=B2*A1
!
FOR J1=1 TO ID
!Delta(1)
FOR J2=1 TO ID
LET DEP(J1,J2,1)=C1(J1,J2)
LET DEM(J1,J2,1)=C2(J1,J2)
LET GAP(J1,J2,1,0)=DEP(J1,J2,1)
LET GAM(J1,J2,1,0)=DEM(J1,J2,1)
NEXT J2
NEXT J1
!
MAT E=Idn(1)      !Identity matrix E
MAT F1=C1*C2
MAT F2=C2*C1
MAT F1=E−F1
MAT F2=E−F2
```

```
MAT D1=F1*G       !V(1)
MAT D2=F2*G
!
MAT A1=INV(D1)      !INV(V(1))
MAT A2=INV(D2)
!
FOR J1=1 TO ID
FOR J2=1 TO ID
LET VP(J1,J2,1)=D1(J1,J2)
LET VM(J1,J2,1)=D2(J1,J2)
NEXT J2
NEXT J1
!******************************************************
!General Step
!
FOR I=2 TO MD !DELTA(I)
FOR J1=1 TO ID
FOR J2=1 TO ID
LET S1=0
LET S2=0
FOR K=0 TO I-2
FOR J3=0 TO ID
LET S1=S1+GAP(J1,J3,I-1,K)*R(J3,J2,K+1)
LET S2=S2+GAM(J1,J3,I-1,K)*R(J2,J3,K+1)
NEXT J3
NEXT K
LET B1(J1,J2)=-R(J1,J2,I)-S1
LET B2(J1,J2)=-R(J2,J1,I)-S2
NEXT J2
NEXT J1
!
MAT C1=B1*A2
MAT C2=B2*A1
!
FOR J1=1 TO ID      ! DELTA(I)
FOR J2=1 TO ID
LET DEP(J1,J2,I)=C1(J1,J2)
LET DEM(J1,J2,I)=C2(J1,J2)
NEXT J2
NEXT J1
!
FOR J1=1 TO ID !GAMMA(I,*)
FOR J2=1 TO ID
LET GAP(J1,J2,I,0)=DEP(J1,J2,I)
LET GAM(J1,J2,I,0)=DEM(J1,J2,I)
```

```
NEXT J2
NEXT J1
!
FOR J1=1 TO ID
FOR J2=1 TO ID
FOR K=1 TO I-1
LET S3=0
LET S4=0
FOR J3=0 TO ID
LET S3=S3+DEP(J1,J3,I)*GAM(J3,J2,I-1,I-1-K)
LET S4=S4+DEM(J1,J3,I)*GAP(J3,J2,I-1,I-1-K)
NEXT J3
LET GAP(J1,J2,I,K)=GAP(J1,J2,I-1,K-1)+S3
LET GAM(J1,J2,I,K)=GAM(J1,J2,I-1,K-1)+S4
NEXT K
NEXT J2
NEXT J1
!
MAT F1=C1*C2
MAT F2=C2*C1
MAT F1=E-F1
MAT F2=E-F2
MAT D1=F1*D1      ! V(I)
MAT D2=F2*D2
!
FOR J1=1 TO ID
FOR J2=1 TO ID
LET VP(J1,J2,I)=D1(J1,J2)
LET VM(J1,J2,I)=D2(J1,J2)
NEXT J2
NEXT J1
MAT A1=INV(D1)
MAT A2=INV(D2)
NEXT I
!***** CAUSAL DIFFRENCE BETWEEN X(1,*) AND X(2,*) *****
LET mean=0
OPEN #3: NAME T3$, ORG TEXT, CREATE NEW, ACCESS OUTPUT
FOR I=1 TO MM
LET mean=mean+V1(I)-VP(1,1,I)
NEXT I
LET mean=mean/MM
PRINT mean
PRINT #3: mean
CLOSE #3
!**************************************************
```

```
PRINT "The calculation finished!"
STOP
END
```

B.2 Calculation of the Local Causality

The programming code is the algorism for calculating the causal values from time sequential data of 1000 random numbers to the time sequential data $\mathcal{X}$. The causal values are stored in the file "dile.dat". The data file "random.dat" of random numbers should be prepared. The programming code is as follows:
Programming Code

```
!Program Name: lcaxu.tru: LOOP-Test of Stationarity
!This program was made by Yuji Nakano
! Causal values for 1000 random numbers will be calculated
DIM A1(2,2),B1(2,2),C1(2,2),D1(2,2),F1(2,2),G(2,2),E(2,2)
DIM A2(2,2),B2(2,2),C2(2,2),D2(2,2),F2(2,2)
DIM T(2,2),TINV(2,2),H(2,1),TH(2,1)
!
OPTION BASE 0
DIM RAN(1000000),RN(1000000),X1(1000)
DIM X(2,1000),Z(2,1000),Z1(2,100),R(2,2,100),TR(2,2,100)
DIM DE1(100),GA1(100,100),V1(100)
DIM DEP(2,2,100),DEM(2,2,100),VP(2,2,100),VM(2,2,100)
DIM GAP(2,2,100,100),GAM(2,2,100,100)
DIM FPZ(2,100),WP(2,100),WZ(100)
!
PRINT "Dimension of data: D=2"
LET ID=2
!
INPUT PROMPT "data number? ": N
LET N=N-1
INPUT PROMPT "1st time sequential data name as ******(.dat)": T1$
LET T1$=T1$ & ".dat"
LET T2$="dile.dat"
! random.dat is the data file of random numbers
OPEN #2: NAME "random.dat", ORG TEXT, CREATE OLD, ACCESS
INPUT
LET NR=(N+1)*1000
FOR I=0 TO NR
INPUT #2: RAN(I)
NEXT I
CLOSE #2
!
```

```
OPEN #1:NAME T1$ ,ORG TEXT,CREATE OLD, ACCESS INPUT
FOR I=0 TO N
INPUT #1:X1(I)
NEXT I
CLOSE #1
!*********** TRANSFORM DATA ************************
!*********** DIFFERENCE ***************
PRINT"D: 0: 0-DIF. 1: 1-DIF. 2: 2-DIF."
!INPUT DIF
LET DIF=0
IF DIF=0 THEN
FOR I=0 TO N
LET X(1,I)=X1(I)
NEXT I
FOR I=0 TO NR
LET RN(I)=RAN(I)
NEXT I
ELSEIF DIF=1 THEN
FOR I=0 TO N-1
LET X(1,I)=X1(I+1)-X1(I)
NEXT I
FOR I=0 TO NR-1
LET RN(I)=RAN(I+1)-RAN(I)
NEXT I
LET N=N-1
LET NR=NR-1
ELSE
FOR I=0 TO N-2
LET X(1,I)=X1(I+2)-2*X1(I+1)+X1(I)
NEXT I
FOR I=0 TO NR-2
LET RN(I)=RAN(I+2)-2*RAN(I+1)+RAN(I)
NEXT I
LET N=N-2
LET NR=NR-2
END IF
!******** LOG TRANS. *************************
PRINT "LOG TRANS.? 0: NO 1: YES "
!INPUT LT
LET LT=0
IF LT=1 THEN
FOR I=0 TO N
LET X(1,I)=LOG(X(1,I))
NEXT I
FOR I=0 TO NR
```

```
LET RAN(I)=LOG(RAN(I))
NEXT I
END IF
!********* ARC TANGENT TRANS. ******************
PRINT "ARC TANGENT TRANS.? 0: NO 1: YES "
!INPUT AT
LET AT=0
IF AT=1 THEN
FOR I=0 TO N
LET X(1,I)=ATN(X(1,I))
NEXT I
FOR I=0 TO NR
LET RAN(I)=ATN(RAN(I))
NEXT I
END IF
!********************************************************
OPEN #3: NAME T2$, CREATE NEWOLD, ACCESS OUTPUT
!********************************************************
FOR IR=1 TO 1000
PRINT IR
!********************************************************
FOR I=0 TO N
LET X(2,I)=RAN((N+1)*(IR−1)+I)
NEXT I
!******************* Normalization of data ****************
FOR J=1 TO ID
LET S=0
FOR I=0 TO N
LET S=S+X(J,I)
NEXT I
LET XBAR=S/(N+1)
FOR I=0 TO N
LET X(J,I)=X(J,I)−XBAR
NEXT I
NEXT J
!
FOR J=1 TO ID       ! Divided by standard deviation
LET SS=0
FOR I=0 TO N
LET SS=SS+X(J,I)*X(J,I)
NEXT I
LET XVAR=SS/(N+1)
FOR I=0 TO N
LET Z(J,I)=X(J,I)/SQR(XVAR)
NEXT I
```

```
NEXT J
!
!*********************************************************
LET M1=INT(3*SQR(N+1))-1
LET MD=INT(2*SQR(N+1)/ID)-1      !Efficient number of R
LET MM=MD
!*********************************************************
! Covariance of normalized data
FOR J1=1 TO ID
FOR J2=1 TO ID
FOR J=0 TO M1
LET CV=0
FOR I=0 TO N-J
LET CV=CV+Z(J1,I+J)*Z(J2,I)
NEXT I
LET R(J1,J2,J)=CV/(N+1) ! Covariance Function
NEXT J
NEXT J2
NEXT J1
!
!************** ALGORITHM : 1-DIM ******************
!****************************************************
! Algorithm of KM2O-Langevin equation
!****************************************************
LET DE1(1)=-R(1,1,1)/R(1,1,0)      !1 STEP
LET GA1(1,0)=DE1(1)
LET V1(0)=R(1,1,0)
LET V1(1)=(1-DE1(1)*DE1(1))*V1(0)
!
LET DE1(2)=-(R(1,1,1)*DE1(1)+R(1,1,2))/V1(1) ! 2 STEP
LET GA1(2,0)=DE1(2)
LET GA1(2,1)=DE1(1)*(1+DE1(2))
LET V1(2)=(1-DE1(2)*DE1(2))*V1(1)
!
FOR I=3 TO M1      ! General Step
LET S1=0
FOR J=0 TO I-2
LET S1=S1+GA1(I-1,J)*R(1,1,J+1)
NEXT J
LET DE1(I)=-(R(1,1,I)+S1)/V1(I-1)
LET V1(I)=(1-DE1(I)*DE1(I))*V1(I-1)
LET GA1(I,0)=DE1(I)
FOR J=1 TO I-1
LET GA1(I,J)=GA1(I-1,J-1)+DE1(I)*GA1(I-1,I-1-J)
NEXT J
```

```
NEXT I
! !********** ALGORITHS : 2-DIM ********************
!**********************************************
! 1-Step
FOR J1=1 TO ID
FOR J2=1 TO ID
LET G(J1,J2)=R(J1,J2,0)
NEXT J2
NEXT J1
!
FOR J1=1 TO ID
FOR J2=1 TO ID
LET VP(J1,J2,0)=G(J1,J2)
LET VM(J1,J2,0)=G(J1,J2)
NEXT J2
NEXT J1
!
MAT A1=INV(G)       !Inverse of R(0)
OR J1=1 TO ID       !R(1)
FOR J2=1 TO ID
LET B1(J1,J2)=−R(J1,J2,1)
LET B2(J1,J2)=−R(J2,J1,1)
NEXT J2
NEXT J1
MAT C1=B1*A1
MAT C2=B2*A1
FOR J1=1 TO ID      !Delta(1)
FOR J2=1 TO ID
LET DEP(J1,J2,1)=C1(J1,J2)
LET DEM(J1,J2,1)=C2(J1,J2)
LET GAP(J1,J2,1,0)=DEP(J1,J2,1)
LET GAM(J1,J2,1,0)=DEM(J1,J2,1)
NEXT J2
NEXT J1
!
MAT E=Idn(1)       !Identity matrix
MAT F1=C1*C2
MAT F2=C2*C1
MAT F1=E−F1
MAT F2=E−F2
MAT D1=F1*G       !V(1)
MAT D2=F2*G
!
MAT A1=INV(D1)      !INV(V(1))
MAT A2=INV(D2)
```

```
!
FOR J1=1 TO ID
FOR J2=1 TO ID
LET VP(J1,J2,1)=D1(J1,J2)
LET VM(J1,J2,1)=D2(J1,J2)
NEXT J2
NEXT J1
!*****************************************************
!General Step
FOR I=2 TO MD       !DELTA(I)
FOR J1=1 TO ID
FOR J2=1 TO ID
LET S1=0
LET S2=0
FOR K=0 TO I-2
FOR J3=0 TO ID
LET S1=S1+GAP(J1,J3,I-1,K)*R(J3,J2,K+1)
LET S2=S2+GAM(J1,J3,I-1,K)*R(J2,J3,K+1)
NEXT J3
NEXT K
LET B1(J1,J2)=-R(J1,J2,I)-S1
LET B2(J1,J2)=-R(J2,J1,I)-S2
NEXT J2
NEXT J1
MAT C1=B1*A2
MAT C2=B2*A1
FOR J1=1 TO ID      !DELTA(I)
FOR J2=1 TO ID
LET DEP(J1,J2,I)=C1(J1,J2)
LET DEM(J1,J2,I)=C2(J1,J2)
NEXT J2
NEXT J1
FOR J1=1 TO ID      !GAMMA(I,*)
FOR J2=1 TO ID
LET GAP(J1,J2,I,0)=DEP(J1,J2,I)
LET GAM(J1,J2,I,0)=DEM(J1,J2,I)
NEXT J2
NEXT J1
FOR J1=1 TO ID
FOR J2=1 TO ID
FOR K=1 TO I-1
LET S3=0
LET S4=0
FOR J3=0 TO ID
LET S3=S3+DEP(J1,J3,I)*GAM(J3,J2,I-1,I-1-K)
```

```
LET S4=S4+DEM(J1,J3,I)*GAP(J3,J2,I−1,I−1−K)
NEXT J3
LET GAP(J1,J2,I,K)=GAP(J1,J2,I−1,K−1)+S3
LET GAM(J1,J2,I,K)=GAM(J1,J2,I−1,K−1)+S4
NEXT K
NEXT J2
NEXT J1
!
MAT F1=C1*C2
MAT F2=C2*C1
MAT F1=E−F1
MAT F2=E−F2
MAT D1=F1*D1      !V(I)
MAT D2=F2*D2
FOR J1=1 TO ID
FOR J2=1 TO ID
LET VP(J1,J2,I)=D1(J1,J2)
LET VM(J1,J2,I)=D2(J1,J2)
NEXT J2
NEXT J1
MAT A1=INV(D1)
MAT A2=INV(D2)
NEXT I
!******************************************************
!******* CAUSAL DIFFRENCE BETWEEN X(1,*) AND X(2,*) ******
LET mean=0
FOR I=1 TO MM
LET mean=mean+V1(I)−VP(1,1,I)
NEXT I
LET mean=mean/MM
!
PRINT #3: mean
!************************************************
NEXT IR
CLOSE #3
!************************************************
PRINT "The calculation finished!!"
STOP
END
```

B.3 Removing the Influence of $\mathcal{Z}$

The programming code is the algorism for removing the influence of the time sequential data $\mathcal{Z}$ from the time sequential data $\mathcal{X}$. The programming code is as follows:

Programming Code

```
! Program name: xyz.tru
!This program was made by Yuji Nakano
! This program extracts a time sequential data z from x
!***********************************************************
! x is affected by z
! y is not affected by z
! The new data is extracted z from x and stored in the file awz
!***********************************************************
DIM F(2,2),G(2,2),P(2,2),Q(2,2)
!
OPTION BASE 0
DIM X1(10000),Y1(10000),Z1(10000)
DIM X(10000),Y(10000),Z(10000),ZR(10000),RA(10000)
DIM RXX(300),RXY(300),RYX(300),RYY(300)
DIM RXZ(300),RZX(300),RZZ(300)
DIM RYZ(300),RZY(300)
DIM DELTAZ(300),GAMMAZ(300,300)
DIM GZ(300,300),VZ(300)
DIM CXZ(300,300),CYZ(300,300)
DIM EYZ(300,300),YNZ(300,300)
DIM EXZ(300,300),XNZ(300,300)
DIM XWZ(10000),NUZ(300)
INPUT PROMPT "Data number ?": N
LET N=N−1
!************** READ DATA ****************************
INPUT PROMPT "Time sequential data name X as ******(.dat)": T1$
INPUT PROMPT "Time sequential data name Y as ******(.dat)": T2$
INPUT PROMPT "Time sequential data name Z as ******(.dat)": T3$
LET T1$=T1 & ".dat"
LET T2$=T2 & ".dat"
LET T3$=T3 & ".dat"
!
! Reading data from data files
OPEN #1:NAME T1$ ,ORG TEXT,CREATE OLD, ACCESS INPUT
OPEN #2:NAME T2$ ,ORG TEXT,CREATE OLD, ACCESS INPUT
OPEN #3:NAME T3$ ,ORG TEXT,CREATE OLD, ACCESS INPUT
!
FOR I=0 TO N
INPUT #1:X1(I)
```

```
NEXT I
CLOSE #1
FOR I=0 TO N
INPUT #2:Y1(I)
NEXT I
CLOSE #2
FOR I=0 TO N
INPUT #3:Z1(I)
NEXT I
CLOSE #3
!*********** TRANSFORM DATA ************************
!*********** DIFFERENCE ****************
!PRINT"D: 0: 0-DIF. 1: 1-DIF. 2: 2-DIF."
!INPUT DIF
LET DIF=0
IF DIF=0 THEN
FOR I=0 TO N
LET X(I)=X1(I)
LET Y(I)=Y1(I)
LET Z(I)=Z1(I)
NEXT I
ELSEIF DIF=1 THEN
FOR I=0 TO N-1
LET X(I)=X1(I+1)-X1(I)
LET Y(I)=Y1(I+1)-Y1(I)
LET Z(I)=Z1(I+1)-Z1(I)
NEXT I
LET N=N-1
ELSE
FOR I=0 TO N-2
LET X(I)=X1(I+2)-2*X1(I+1)+X1(I)
LET Y(I)=Y1(I+2)-2*Y1(I+1)+Y1(I)
LET Z(I)=Z1(I+2)-2*Z1(I+1)+Z1(I)
NEXT I
LET N=N-2
END IF
!******** LOG TRANS. **********************
!PRINT "LOG TRANS.? 0: NO 1: YES "
!INPUT LT
LET LT=0
IF LT=1 THEN
FOR I=0 TO N
LET X(I)=LOG(X(I))
LET Y(I)=LOG(Y(I))
LET Z(I)=LOG(Z(I))
```

```
NEXT I
END IF
!********* ARC TANGENT TRANS. *****************
!PRINT "ARC TANGENT TRANS.? 0: NO 1: YES "
!INPUT AT
LET AT=0
IF AT=1 THEN
FOR I=0 TO N
LET X(I)=ATN(X(I))
LET Y(I)=ATN(Y(I))
LET Z(I)=ATN(Z(I))
NEXT I
END IF
!**********************************************
!******* Normalization of data *****************
LET XBAR=0
LET YBAR=0
LET ZBAR=0
FOR I=0 TO N
LET XBAR=XBAR+X(I)
LET YBAR=YBAR+Y(I)
LET ZBAR=ZBAR+Z(I)
NEXT I
LET XBAR=XBAR/(N+1)
LET YBAR=YBAR/(N+1)
LET ZBAR=ZBAR/(N+1)
FOR I=0 TO N
LET X(I)=X(I)-XBAR
LET Y(I)=Y(I)-YBAR
LET Z(I)=Z(I)-ZBAR
NEXT I
LET S1=0
LET S2=0
LET S3=0
FOR I=0 TO N
LET S1=S1+X(I)*X(I)
LET S2=S2+Y(I)*Y(I)
LET S3=S3+Z(I)*Z(I)
NEXT I
LET XVAR=S1/(N+1)        ! The variance of data
LET YVAR=S2/(N+1)
LET ZVAR=S3/(N+1)
! Normalization of data
FOR I=0 TO N
LET X(I)=X(I)/SQR(XVAR)
```

```
LET Y(I)=Y(I)/SQR(YVAR)
LET Z(I)=Z(I)/SQR(ZVAR)
NEXT I
!*********************************************
!The largest efficient number
!*********************************************
LET M1=INT(3*SQR(N+1))−1
LET M2=INT(2*SQR(N+1)/2)−1
LET M3=INT(3*(SQR(N+1))/3)−1
!*********************************************
FOR J=0 TO M1
LET R11=0
LET R12=0
LET R13=0
LET R21=0
LET R22=0
LET R23=0
LET R31=0
LET R32=0
LET R33=0
FOR I=0 TO N−J
LET R11=R11+X(I+J)*X(I)
LET R12=R12+X(I+J)*Y(I)
LET R13=R13+X(I+J)*Z(I)
LET R21=R21+Y(I+J)*X(I)
LET R22=R22+Y(I+J)*Y(I)
LET R23=R23+Y(I+J)*Z(I)
LET R31=R31+Z(I+J)*X(I)
LET R32=R32+Z(I+J)*Y(I)
LET R33=R33+Z(I+J)*Z(I)
NEXT I
LET RXX(J)=R11/(N+1)
LET RXY(J)=R12/(N+1)
LET RXZ(J)=R13/(N+1)
LET RYX(J)=R21/(N+1)
LET RYY(J)=R22/(N+1)
LET RYZ(J)=R23/(N+1)
LET RZX(J)=R31/(N+1)
LET RZY(J)=R32/(N+1)
LET RZZ(J)=R33/(N+1)
NEXT J
!************* ALGOLISM *****************
! Algorithm of KM2O-Langevin equation
LET DELTAZ(1)=−RZZ(1)/RZZ(0)      !1 STEP
LET GZ(1,0)=DELTAZ(1)
```

```
LET VZ(0)=RZZ(0)
LET VZ(1)=(1−DELTAZ(1)*DELTAZ(1))*VZ(0)
!
LET DELTAZ(2)=−(RZZ(1)*DELTAZ(1)+RZZ(2))/VZ(1)      ! 2 STEP
LET GZ(2,0)=DELTAZ(2)
LET GZ(2,1)=DELTAZ(1)*(1+DELTAZ(2))
LET VZ(2)=(1−DELTAZ(2)*DELTAZ(2))*VZ(1)
!
FOR I=3 TO M1      ! General Step
LET S1=0
FOR J=0 TO I−2
LET S1=S1+GZ(I−1,J)*RZZ(J+1)
NEXT J
LET DELTAZ(I)=−(RZZ(I)+S1)/VZ(I-1)
LET VZ(I)=(1−DELTAZ(I)*DELTAZ(I))*VZ(I−1)
LET GZ(I,0)=DELTAZ(I)
FOR J=1 TO I−1
LET GZ(I,J)=GZ(I−1,J−1)+DELTAZ(I)*GZ(I−1,I−1−J)
NEXT J
NEXT I
!****************************************
!    COMPUTING CXZ(*,*)
!****************************************
LET CXZ(0,0)=RXZ(0)/RZZ(0)
FOR I=1 TO M2
LET CXZ(I,0)=RXZ(I)/RZZ(0)
FOR J=1 TO I
LET S1=0
FOR L=0 TO J−1
LET S1=S1+GZ(J,L)*RXZ(I−L)
NEXT L
LET CXZ(I,J)=(RXZ(I−J)+S1)/VZ(J)
NEXT J
NEXT I
!************************************
!    (X(*),NUZ(**))
!************************************
!**** (X(I),NUZ(K)) I>=K *************
FOR I=0 TO M2
LET XNZ(I,0)=RXZ(I)
NEXT I
FOR I=1 TO M2
FOR K=1 TO I
LET S1=0
FOR L=0 TO K−1
```

```
LET S1=S1+GZ(K,L)*RXZ(I-L)
NEXT L
LET XNZ(I,K)=RXZ(I-K)+S1
NEXT K
NEXT I
!********** XWZ(*) *********************
LET XWZ(0)=X(0)
FOR I=1 TO M2
LET NUZ(0)=Z(0)
FOR K=1 TO I
LET S=0
FOR J=0 TO K-1
LET S=S+GZ(K,J)*Z(J)
NEXT J
LET NUZ(K)=Z(K)+S
NEXT K
LET S1=0
FOR J=0 TO I-1
LET S1=S1+CXZ(I,J)*NUZ(J)
NEXT J
LET XWZ(I)=X(I)-S1
NEXT I
!
FOR I=M2+1 TO N
LET NUZ(0)=Z(I-M2)
FOR K=1 TO M2
LET S2=0
FOR J=0 TO K-1
LET S2=S2+GZ(K,J)*Z(I-M2+J)
NEXT J
LET NUZ(K)=Z(I-M2+K)+S2
NEXT K
LET S3=0
FOR J=0 TO M2-1
LET S3=S3+CXZ(M2,J)*NUZ(J)
NEXT J
LET XWZ(I)=X(I)-S3
NEXT I
!*******************************************************
OPEN #4: NAME awz$, CREATE NEWOLD, ACCESS OUTPUT
OPEN #5: NAME bbb$, CREATE NEWOLD, ACCESS OUTPUT
!*******************************************************
FOR I=0 TO N
PRINT #4:XWZ(I)
NEXT I
```

```
CLOSE #4
!
FOR I=0 TO N
PRINT #5:Y(I)
NEXT I
CLOSE #5
!**************************************************
PRINT "The calculation finished!!"
PRINT "bbb is dummy file and should be deleted!!"
STOP
END
```

B.4 Sorting 1000 Causal Values

The programming code is the algorism for sorting the 1000 causal values of the time sequential data **X** due to time sequential random numbers in the ascending order, and seeking the percentile of the causal value of the time sequential data **X** due to the time sequential data **Y**. The programming code is as follows:

Programming Code

```
!Program Name: "sort&seek.tru"
!This program was made by Osamu Morita
!Program for sorting the 1000 causal values in ascending order,
!and finding the percentile of the causal value.
DIM f1(1000),g1(1000),g2(1000),h1(1000),h2(1000),h3(1000),h4(1000)
LET dn=1000
LET gm$="dile.dat" ! Data file of 1000 causal value
LET gn$="dilew.dat" ! Data file of sorted 1000 causal values
INPUT PROMPT "File name of the causal value as ******(.dat)": go$
LET go$=go$ & ".dat"
LET gp$="PT.dat" ! Data file of the percentile of the causal value
!*********************************************************
!Following 14 files are tentative and should be deleted
! after the calculation finished.
LET Aa$="Aa.dat"
LET Ab$="Ab.dat"
LET Ac$="Ac.dat"
LET Ad$="Ad.dat"
LET Ae$="Ae.dat"
LET Af$="Af.dat"
LET w1$="w1.dat"
LET w2$="w2.dat"
LET w3$="w3.dat"
```

```
LET w4$="w4.dat"
LET w5$="w5.dat"
LET w6$="w6.dat"
LET w7$="w7.dat"
LET w8$="w8.dat"
!********************************************************
!Reading data from data files
OPEN #1: NAME gm$, ORG TEXT, CREATE OLD, ACCESS INPUT
FOR i=1 TO dn
INPUT #1: f1(i)
NEXT i
CLOSE #1
! 1-st step: dividing the original file to 2 sub-files
LET sum=0.0
FOR i=1 TO dn
LET sum=sum+f1(i)
NEXT i
LET vm0=sum/dn
OPEN #2: NAME Aa$, ORG TEXT, CREATE NEWOLD, ACCESS OUT-
PUT
OPEN #3: NAME Ab$, ORG TEXT, CREATE NEWOLD, ACCESS OUT-
PUT
LET na=0
LET nb=0
FOR i=1 TO dn
IF f1(i) <= vm0 THEN
PRINT #2: f1(i)
LET na=na+1
ELSEIF f1(i) > vm0 THEN
PRINT #3: f1(i)
LET nb=nb+1
END IF
NEXT i
CLOSE #2
CLOSE #3
OPEN #2: NAME Aa$, ORG TEXT, CREATE NEWOLD, ACCESS INPUT
FOR i=1 TO na
INPUT #2: g1(i)
NEXT i
CLOSE #2
OPEN #3: NAME Ab$, ORG TEXT, CREATE NEWOLD, ACCESS INPUT
FOR i=1 TO nb
INPUT #3: g2(i)
NEXT i CLOSE #3
! 2-nd step: dividing the original file to 4 sub-files
```

```
LET sum=0.0
FOR i=1 TO na
LET sum=sum+g1(i)
NEXT i
LET vm21=sum/na
LET sum=0.0
FOR i=1 TO nb
LET sum=sum+g2(i)
NEXT i
LET vm22=sum/nb
OPEN #1: NAME Ac$, ORG TEXT, CREATE NEWOLD, ACCESS OUT-
PUT
OPEN #2: NAME Ad$, ORG TEXT, CREATE NEWOLD, ACCESS OUT-
PUT
OPEN #3: NAME Ae$, ORG TEXT, CREATE NEWOLD, ACCESS OUT-
PUT
OPEN #4: NAME Af$, ORG TEXT, CREATE NEWOLD, ACCESS OUT-
PUT
LET nc=0
LET nd=0
FOR i=1 TO na
IF g1(i) <= vm21 THEN
PRINT #1: g1(i)
LET nc=nc+1
ELSEIF g1(i) > vm21 THEN
PRINT #2: g1(i)
LET nd=nd+1
END IF
NEXT i
CLOSE #1
CLOSE #2
LET ne=0
LET nf=0
FOR i=1 TO nb
IF g2(i) <= vm22 THEN
PRINT #3: g2(i)
LET ne=ne+1
ELSEIF g2(i) > vm22 THEN
PRINT #4: g2(i)
LET nf=nf+1
END IF
NEXT i
CLOSE #3
CLOSE #4
! 3-rd step: dividing the original file to 8 sub-files
```

```
OPEN #1: NAME Ac$, ORG TEXT, CREATE NEWOLD, ACCESS INPUT
FOR i=1 TO nc
INPUT #1: h1(i)
NEXT i
CLOSE #1
OPEN #2: NAME Ad$, ORG TEXT, CREATE NEWOLD, ACCESS INPUT
FOR i=1 TO nd
INPUT #2: h2(i)
NEXT i
CLOSE #2
OPEN #3: NAME Ae$, ORG TEXT, CREATE NEWOLD, ACCESS INPUT
FOR i=1 TO ne
INPUT #3: h3(i)
NEXT i
CLOSE #3
OPEN #4: NAME Af$, ORG TEXT, CREATE NEWOLD, ACCESS INPUT
FOR i=1 TO nf
INPUT #4: h4(i)
NEXT i
CLOSE #4
LET sum=0.0
FOR i=1 TO nc
LET sum=sum+h1(i)
NEXT i
LET vm31=sum/nc
LET sum=0.0
FOR i=1 TO nd
LET sum=sum+h2(i)
NEXT i
LET vm32=sum/nd
LET sum=0.0
FOR i=1 TO ne
LET sum=sum+h3(i)
NEXT i
LET vm33=sum/ne
LET sum=0.0
FOR i=1 TO nf
LET sum=sum+h4(i)
NEXT i
LET vm34=sum/nf
! 3-1 step:
OPEN #1: NAME w1$, ORG TEXT, CREATE NEWOLD, ACCESS OUT-
PUT
OPEN #2: NAME w2$, ORG TEXT, CREATE NEWOLD, ACCESS OUT-
PUT
```

```
LET n1=0
LET n2=0
FOR i=1 TO nc
IF h1(i) <= vm31 THEN
PRINT #1: h1(i)
LET n1=n1+1
ELSEIF h1(i) > vm31 THEN
PRINT #2: h1(i)
LET n2=n2+1
END IF
NEXT i
CLOSE #1
CLOSE #2
! 3-2 step:
OPEN #1: NAME w3$, ORG TEXT, CREATE NEWOLD, ACCESS OUT-
PUT
OPEN #2: NAME w4$, ORG TEXT, CREATE NEWOLD, ACCESS OUT-
PUT
LET n3=0
LET n4=0
FOR i=1 TO nd
IF h2(i) <= vm32 THEN
PRINT #1: h2(i)
LET n3=n3+1
ELSEIF h2(i) > vm32 THEN
PRINT #2: h2(i)
LET n4=n4+1
END IF
NEXT i
CLOSE #1
CLOSE #2
! 3-3 step:
OPEN #1: NAME w5$, ORG TEXT, CREATE NEWOLD, ACCESS OUT-
PUT
OPEN #2: NAME w6$, ORG TEXT, CREATE NEWOLD, ACCESS OUT-
PUT
LET n5=0
LET n6=0
FOR i=1 TO ne
IF h3(i) <= vm33 THEN
PRINT #1: h3(i)
LET n5=n5+1
ELSEIF h3(i) > vm33 THEN
PRINT #2: h3(i)
LET n6=n6+1
```

```
END IF
NEXT i
CLOSE #1
CLOSE #2
! 3-4 step:
OPEN #1: NAME w7$, ORG TEXT, CREATE NEWOLD, ACCESS OUT-
PUT
OPEN #2: NAME w8$, ORG TEXT, CREATE NEWOLD, ACCESS OUT-
PUT
LET n7=0
LET n8=0
FOR i=1 TO nf
IF h4(i) <= vm34 THEN
PRINT #1: h4(i)
LET n7=n7+1
ELSEIF h4(i) > vm34 THEN
PRINT #2: h4(i)
LET n8=n8+1
END IF
NEXT i
CLOSE #1
CLOSE #2
! Sorting of 8 sub-files
! 1-st step
OPEN #1: NAME w1$, ORG TEXT, CREATE OLD, ACCESS INPUT
FOR i=1 TO n1
INPUT #1: f1(i)
NEXT i
CLOSE #1
FOR k=1 TO n1
LET vmin=1.0
FOR i=1 TO n1
LET vmin=MIN(vmin,f1(i))
NEXT i
FOR i=1 TO n1
IF f1(i)=vmin THEN EXIT FOR
NEXT i
LET g1(k)=f1(i)
LET f1(i)=10.0
NEXT k
! 2-nd step
OPEN #1: NAME w2$, ORG TEXT, CREATE OLD, ACCESS INPUT
FOR i=1 TO n2
INPUT #1: f1(i)
NEXT i
```

```
CLOSE #1
LET n11=n1+1
LET n12=n1+n2
FOR k=n11 TO n12
LET vmin=1.0
FOR i=1 TO n2
LET vmin=MIN(vmin,f1(i))
NEXT i
FOR i=1 TO n2
IF f1(i)=vmin THEN EXIT FOR
NEXT i
LET g1(k)=f1(i)
LET f1(i)=10.0
NEXT k
! 3-rd step
OPEN #1: NAME w3$, ORG TEXT, CREATE OLD, ACCESS INPUT
FOR i=1 TO n3
INPUT #1: f1(i)
NEXT i
CLOSE #1
LET n21=n1+n2+1
LET n22=n1+n2+n3
FOR k=n21 TO n22
LET vmin=1.0
FOR i=1 TO n3
LET vmin=MIN(vmin,f1(i))
NEXT i
FOR i=1 TO n3
IF f1(i)=vmin THEN EXIT FOR
NEXT i
LET g1(k)=f1(i)
LET f1(i)=10.0
NEXT k
! 4-th step
OPEN #1: NAME w4$, ORG TEXT, CREATE OLD, ACCESS INPUT
FOR i=1 TO n4
INPUT #1: f1(i)
NEXT i
CLOSE #1
LET n31=n1+n2+n3+1
LET n32=n1+n2+n3+n4
FOR k=n31 TO n32
LET vmin=1.0
FOR i=1 TO n4
LET vmin=MIN(vmin,f1(i))
```

```
NEXT i
FOR i=1 TO n4
IF f1(i)=vmin THEN EXIT FOR
NEXT i
LET g1(k)=f1(i)
LET f1(i)=10.0
NEXT k
! 5-th step
OPEN #1: NAME w5$, ORG TEXT, CREATE OLD, ACCESS INPUT
FOR i=1 TO n5
INPUT #1: f1(i)
NEXT i
CLOSE #1
LET n41=n1+n2+n3+n4+1
LET n42=n1+n2+n3+n4+n5
FOR k=n41 TO n42
LET vmin=1.0
FOR i=1 TO n5
LET vmin=MIN(vmin,f1(i))
NEXT i
FOR i=1 TO n5
IF f1(i)=vmin THEN EXIT FOR
NEXT i
LET g1(k)=f1(i)
LET f1(i)=10.0
NEXT k
! 6-th step
OPEN #1: NAME w6$, ORG TEXT, CREATE OLD, ACCESS INPUT
FOR i=1 TO n6
INPUT #1: f1(i)
NEXT i
CLOSE #1
LET n51=n1+n2+n3+n4+n5+1
LET n52=n1+n2+n3+n4+n5+n6
FOR k=n51 TO n52
LET vmin=1.0
FOR i=1 TO n6
LET vmin=MIN(vmin,f1(i))
NEXT i
FOR i=1 TO n6
IF f1(i)=vmin THEN EXIT FOR
NEXT i
LET g1(k)=f1(i)
LET f1(i)=10.0
NEXT k
```

```
! 7-th step
OPEN #1: NAME w7$, ORG TEXT, CREATE OLD, ACCESS INPUT
FOR i=1 TO n7
INPUT #1: f1(i)
NEXT i
CLOSE #1
LET n61=n1+n2+n3+n4+n5+n6+1
LET n62=n1+n2+n3+n4+n5+n6+n7
FOR k=n61 TO n62
LET vmin=1.0
FOR i=1 TO n7
LET vmin=MIN(vmin,f1(i))
NEXT i
FOR i=1 TO n7
IF f1(i)=vmin THEN EXIT FOR
NEXT i
LET g1(k)=f1(i)
LET f1(i)=10.0
NEXT k
! 8-th step
OPEN #1: NAME w8$, ORG TEXT, CREATE OLD, ACCESS INPUT
FOR i=1 TO n8
INPUT #1: f1(i)
NEXT i
CLOSE #1
LET n71=n1+n2+n3+n4+n5+n6+n7+1
LET n72=n1+n2+n3+n4+n5+n6+n7+n8
FOR k=n71 TO n72
LET vmin=1.0
FOR i=1 TO n8
LET vmin=MIN(vmin,f1(i))
NEXT i
FOR i=1 TO n8
IF f1(i)=vmin THEN EXIT FOR
NEXT i
LET g1(k)=f1(i)
LET f1(i)=10.0
NEXT k
OPEN #2: NAME gn$, ORG TEXT, CREATE NEW, ACCESS OUTPUT
FOR i=1 TO dn
PRINT #2: g1(i)
NEXT i
CLOSE #2
!*******************************************************
! Seeking the percentile of the causal value!
```

```
!***********************************************************
OPEN #1: NAME gn$, ORG TEXT, CREATE OLD, ACCESS INPUT
FOR i=1 TO dn
INPUT #1: f1(i)
NEXT i
CLOSE #1
OPEN #2: NAME go$, ORG TEXT, CREATE OLD, ACCESS INPUT
INPUT #2: cv
CLOSE #2
!PRINT cv
OPEN #1: NAME gp$, ORG TEXT, CREATE NEW, ACCESS OUTPUT
IF cv < f1(1) THEN
LET pt=0.1
PRINT #1: pt
ELSEIF cv > f1(1000) THEN
LET pt=100.0
PRINT #1: pt
ELSE
!
IF cv > f1(1) and cv <= vm31 THEN
LET is=1
LET ie=n1
ELSEIF cv > vm31 and cv <= vm21 THEN
LET is=n1+1
LET ie=n1+n2
ELSEIF cv > vm21 and cv <= vm32 THEN
LET is=n1+n2+1
LET ie=n1+n2+n3
ELSEIF cv > vm32 and cv <= vm0 THEN
LET is=n1+n2+n3+1
LET ie=n1+n2+n3+n4
ELSEIF cv > vm0 and cv <= vm33 THEN
LET is=n1+n2+n3+n4+1
LET ie=n1+n2+n3+n4+n5
ELSEIF cv > vm33 and cv <= vm22 THEN
LET is=n1+n2+n3+n4+n5+1
LET ie=n1+n2+n3+n4+n5+n6
ELSEIF cv > vm22 and cv <= vm34 THEN
LET is=n1+n2+n3+n4+n5+n6+1
LET ie=n1+n2+n3+n4+n5+n6+n7
ELSEIF cv > vm34 and cv <= f1(1000) THEN
LET is=n1+n2+n3+n4+n5+n6+n7+1
LET ie=dn
END IF
!
```

```
FOR i=is TO ie
IF cv > f1(i) and cv <= f1(i+1) THEN EXIT FOR
NEXT i
LET pt=0.1*i
PRINT #1: pt
END IF
CLOSE #1
PRINT "The calculation finished!!"
STOP
END
```

Bibliography

[1] Allan, R.J., N. Nicholls, P.D. Jones, and I.J. Butterworth (1991): A further extension of the Tahiti–Darwin SOI, early SOI results and Darwin pressure, *J. Climate* **4**, 743–749.

[2] Angell, J.K., and J. Korshover (1983a): Global temperature variations in the troposphere and stratosphere, *Mon. Wea. Rev.*, **111**, 901–921.

[3] Angell, J.K., and J. Korshover (1983b): Comparison of stratospheric warmings following Agung and Chichón, *Mon. Wea. Rev.*, **111**, 2129–21351.

[4] Angell, J.K. (1996): Stratospheric temperature after volcanic eruption, In *The Mount Pinatubo Eruption: Effects on the Atmosphere and Climate*, eds. G. Fiocco et al., Springer, Berlin, pp.83–93.

[5] Bluth, G.J.S., S.D. Doiron, C.C. Schneider, A.J. Krueger, and L.S. Walter (1992): Global tracking of the SO_2 clouds from the June, 1991 Mount Pinatubo eruptions, *Geophys. Res. Lett.*, **19**, 151–154.

[6] Bronninmann, S., J. Luterbacher, J. Staehelin, T.M. Svendby, G. Hansen, and T. Svenœ(2004): Extreme climate of the global troposphere and stratosphere 1940–1942 related to El Niño, *Nature*, **431**, 971–974.

[7] Cassou, C. and L. Terray (2001): Dual influence of Atlantic and Pacific SST anomalies on the North Atlantic/Europe winter climate, *Geophys. Res. Lett.*, **28**, 3195–3198.

[8] Fiocco, G., M. Cacciani, A.D. di Sarra, D. Fuà, P. Colagrande, G. De Benedetti, P. di Girolano, R. Viola, D. di Fisica (1996): The evolution of the Pinatubo aerosol layer observed by Lidar at South Pole, Rome, Thule: a summary of the results, In *The Mount Pinatubo Eruption: Effects on the Atmosphere and Climate*, eds. G. Fiocco et al., Springer, Berlin, pp.127–139.

[9] Freadrich, K. and K. Müller (1992): Climate anomalies in Europe associated with ENSO extremes, *Int. J. Climatol.*, **12**, 25–31.

[10] Fujita, T. (1985): The abnormal temperature rises in the lower stratosphere after the 1982 eruption of the volcano El Chichón, Mexico, *Papers Meteor. and Geophys.*, **56**, 495–507.

[11] Gill, A.E. (1980): Some simple solutions for heat-induced tropical circulation, *Quart. J. Roy. Meteor. Soc.*, **106**(449), 447–462.

[12] Granger C.W.J. (1969): Investigating causal relations by econometric models and cross-spectral methods, *Econometrica*, **37** No.3, 424–438.

[13] Held, I.M., M. Ting, and H. Wang (2002): Northern Winter Stationary Waves: Theory and Modeling, *J. Climate*, **15**(16), 2125–2144.

[14] Huang, J., K. Higuchi, and A. Shabbar (1998): The relationship between the North Atlantic Oscillation and the El Niño-Southern Oscillation, *Geophys. Res. Lett.*, **25**, 2707–2710.

[15] Hurrell, J.W. (1995): Decadal trends in the North Atlantic Oscillation: Regional temperatures and precipitation, *Science*, **269**, 676–679.

[16] Hurrell, J.W., Y. Kushnir, G. Ottersen, and M. Visbeck (2003): An Overview of the North Atlantic Oscillation, In *The North Atlantic Oscillation: Climatic Significance and Environmental Impact*, Hurrell, J.W., Y. Kushnir, G. Ottersen, and M. Visbeck eds., *American Geophysical Union, Geophysical Monograph Series*, **134**, p279, ISBN:9781118669037, DOI:10.1029/GM134, pp.1–35.

[17] Jones, P.D., and P.M. Kelly (1996): The effect of tropical explosive volcanic eruptions on surface air temperature, In *The Mount Pinatubo Eruption: Effects on the Atmosphere and Climate*, eds. G. Fiocco et.al., Springer, Berlin, pp.95–111.

[18] Jones, P.D., T. Jónsson, and D. Wheeler (1997): Extension to the North Atlantic Oscillation using early instrumental pressure observations from Gibraltar and South-West Iceland, *Int. J. Climatol.* **17**, 1433–1450.

[19] Jones, P.D., T.J. Osborn, and K.R. Briffa (1997): Estimating sampling errors in large-scale temperature averages, *J. Climate*, **10**, 2548–2568.

[20] Jones, P.D., M. New, D.E. Parker, S. Martin, and I.G. Rigor (1999): Surface air temperature and its changes over the past 150 years, *Rev. Geophys.*, **37**, 173–199.

[21] Jones, P.D., T.J. Osborn, K.R. Briffa, C.K. Folland, B. Horton, L.V. Alexander, D.E. Parker, and N. Rayner (2001): According for sampling density in grid-box land and ocean surface temperature time series, *J. Geophys. Res.*, **106**, 3371–3380.

[22] Jones, P.D., and A. Moberg (2003): Hemispheric and large-scale air temperature variations: An extensive revision and an update to 2001, *J. Climate,* **16**, 206–223.

[23] Jones, P.D. T.J. Osborn, and K.R. Briffa (2003): Pressure-Based Measures of the North Atlantic Oscillations (NAO): A Comparison and an Assessment of Changes in the Strength of the NAO and in its Influence on Surface Climate Parameters, In *The North Atlantic Oscillation: Climatic Significance and Environmental Impact*, Hurrell, J.W., Y. Kushnir, G. Ottersen, and M. Visbeck eds., *American Geophysical Union, Geophysical Monograph Series*, **134**, p279, ISBN:9781118669037, DOI:10.1029/GM134, pp.51–62.

[24] Kalnay, E., M. Kanamitsu, R. Kistler, W. Collins, D. Deaven, L. Gandin, M. Iredell, S. Saha, G. White, J. Woollen, Y. Zhu, M. Chelliah, W. Ebisuzaki, W. Higgins, J. Janowiak, K. C. Mo, C. Ropelewski, J. Wang, A. Leetmaa, R. Reynolds, R. Jenne, and D. Joseph (1996): The NCEP/NCAR 40–year reanalysis project, *Bull. Amer. Meteor. Soc.*, **77**, 437-472.

[25] Kelly, P.M., and C.B. Sear (1984): Climatic impact of explosive volcanic explosions, *Nature*, **311**, 740–743.

[26] Klimek, M., M. Matsuura, and Y. Okabe (2008): Stochastic flows and finite block frames, *J. Math. Analy. and Appl.*, **342**, 816–829.

[27] Können, G.P., P.D. Jones, M.H. Kaltofen, and R.J. Allan (1998): Pre-1866 extensions of the Southern Oscillation Index using early Indonesian and Tahitian meteorological readings, *J. Climate*, **11**, 2325–2339.

[28] Labitzke, K., B. Naujokat, and M.P. McCormick (1983): Temperature effects on the stratosphere of the April 4, 1982 eruption of El Chichón, Mexico, *Geophys. Res. Lett.*, **10**, 24–26.

[29] Labitzke, K., and H. van Loon (1996): The effect on the stratosphere of three tropical volcanic eruptions, In *The Mount Pinatubo Eruption: Effects on the Atmosphere and Climate*, eds. G. Fiocco et al., Springer, Berlin, pp.113–125.

[30] Lamb, H.H. (1970): Volcanic dust in the atmosphere, with a chronology and assessment of its meteorological significance, *Philos. Trans. R. Soc. London*, **A, 266**, 425–533.

[31] Lamb, H.H. (1977): Supplementary volcanic dust veil index assessments, *Clim. Monit.*, **6**, 57–67.

[32] Lamb, H.H. (1983): Update of the chronology of assessments of the volcanic dust veil index, *Clim. Monit.*, **12**, 79–90.

[33] Maeda, S. (2013): The ENSO and the climate of Japan, In *The Latest Study of the El Niño and Southern Oscillation, Study Note of Meteorology*, **228**, eds. Watanabe, M and M. Kimoto, Meteorological Society of Japan, Tokyo, pp.167–179, (*in Japanese*)

[34] McCormick, M.P., and R.E. Veiga (1992): SAGE II measurements of early Pinatubo aerosols, *Geophys. Res. Lett.*, **19**, 155–158.

[35] Michell, J.M. Jr. (1992): Recent secular changes of the global temperature, *Ann. N.Y. Acad. Sci.*, **95**, 235–250.

[36] Nakano, Y. (1995): On a causal analysis of economic time series, *Hokkaido Math. J.*, **23**, 179–213.

[37] Nakano, Y. (1999): On a local causal value of time series, *Working paper of Shiga Univ.*, **63**, 1–11.

[38] Nakano, Y. (1999): On a local causal analysis of time series, *Fourth International Congress on Industrial and Applied Mathematics, Book of Abstract*, Edinburgh Press, pp292.

[39] Nakano, Y. and Y. Okabe (2012): A time series analysis of economical phenomena in Japan's lost decade (I): determinacy property of the velocity of money and equilibrium solution, *Asia-Pacific Financial Markets*, Springer, **19**, pp371–389.

[40] Okabe, Y. (1988): On stochastic difference equations for the multi-dimensional weakly stationary time series, *Prospect of Algebraic Analysis* (eds. Kashiwara M. and T. Kawai), Academic Press, **2**, pp601–645.

[41] Okabe, Y. and Y. Nakano (1991): The theory of KM_2O-Langevin equations and its applications to data analysis (I): Stationary analysis, *Hokkaido Math. J.*, **20**, 45–90.

[42] Okabe, Y. and A. Inoue (1994): The theory of KM_2O-Langevin equations and its applications to data analysis (II): Causal analysis, *Nagoya Math. J.*, **134**, 1–28.

[43] Okabe, Y. and T. Yamane (1998): The theory of KM_2O-Langevin equations and its applications to data analysis (III): Deterministic Analysis, *Nagoya Math. J.*, **152**, 175–201.

[44] Osborn, T.J. (2011): Winter 2009/2010 Temperatures and a Record–Breaking North Atlantic Oscillation Index, *Weather*, **66**, 19–21.

[45] Parker, D.E., and J.L. Branscombe (1983): Stratospheric warming following the El Chichón eruption, *Nature*, **301**, 406–408.

[46] Quadrelli, R. and J.M. Wallace (2002): Dependence of the structure of the Northern Hemisphere annular mode on the polarity of ENSO, *Geophys. Res. Lett.*, **29**, 2132, DOI:10.1029/2002GL015807.

[47] Ramadan, H.H., A.S. Ramamurthy, and R.E. Beighley (2011): Interannual temperature and precipitation variations over the Litani Basin in response to atmospheric circulation patterns, *Theor. Appl. Climatology*. **108**(3-4), 563–577.

[48] Reynolds, R.W., N.A. Rayner, T.M. Smith, D.C. Stokes, and W. Wang (2002): An improved in situ and satellite SST analysis for climate, *J. Climate*, **15**, 1609–1625.

[49] Robock, A., and J. Mao (1992): Winter warming from large volcanic eruptions, *Geophy. Res. Lett.*, **19**, 2405–2408.

[50] Robock, A., and J. Mao (1995): The volcanic signal in surface temperature observations, *J. Climate*, **8**, 1086–1103.

[51] Ropelewski, C.F. and P.D. Jones (1987): An extension of the Tahiti–Darwin Southern Oscillation Index, *Mon. Wea. Rev.*, **115**, 2161–2165.

[52] Schneider, D.P., C. Deser, J. Fasullo, and K. E. Trenberth (2013): Climate Data Guide Spurs Discovery and Understanding, *Eos Trans. AGU*, **94**, 121–122.

[53] Seager, R., N. Harnik, Y. Kushnir, W. Robinson, and J. Miller (2003): Mechanisms of Hemispherically Symmetric Climate Variability, *J. Climate*, **16**(18), 2960–2978.

[54] Seager, R., N. Harnik, W.A. Robinson, Y. Kushnir, M. Ting, H.P. Huang, and J. Velez (2005): Mechanisms of ENSO-forcing of hemispherically symmetric precipitation variability, *Quart. J. Roy. Meteor. Soc.* **131**(608), 1501–1527.

[55] Sear, C.B., P.M. Kelly, P.D. Jones, and C.M. Goodes (1987): Global surface-temperature responses to major volcanic explosions, *Nature*, **330**, 365–367.

[56] Simmons, A.J., J.M. Wallace, and G.W. Branstator (1983): Barotropic Wave Propagation and Instability, and Atmospheric Teleconnection Patterns, *J. Atmos. Sci.*, **40**(6), 1363–1392.

[57] Walker, G.T. (1923): Correlation in seasonal variations of weather, VIII. A preliminary study of world weather, *Memoirs of the India Meteorological Department*, **24**(4), 75–131.

[58] Walker, G.T. (1924): Correlation in seasonal variations of weather, IX. A further study of world weather, *Memoirs of the India Meteorological Department*, **24**(9), 275–3331.

[59] Xie, P., and P.A. Arkin (1997): Global precipitation: A 17-year monthly analysis based on gauge observations, satellite estimates, and numerical model outputs, *Bull. Amer. Meteor. Soc.*, **78**, 2539–2558.

Index

Printed in the United States
by Baker & Taylor Publisher Services